Vienna Circle Institute Library

Volume 6

This peer-reviewed series includes monographs and edited volumes, which complement the format of the related series "Vienna Circle Institute Yearbook" and "Veröffentlichungen des Institut Wiener Kreis", both published with Springer Nature. The books mainly deal with individual members of the Vienna Circle, the entire Schlick-Circle as a collective, its adherents and critics, as well as with related topics of Logical Empiricism and its periphery in historical and philosophical perspective. Specifically, they are based on so far unpublished primary sources and feature also forgotten and marginalized issues as significant contributions to the most recent research in these fields.

More information about this series at https://link.springer.com/bookseries/7041

Jan von Plato

Chapters from Gödel's Unfinished Book on Foundational Research in Mathematics

Vienna
Circle
Institute
Library

Springer

Jan von Plato
Department of Philosophy
University of Helsinki
Helsinki, Finland

The preparation of this book has been financed by the European Research Council Advanced Grant GODELIANA (grant agreement No 787758).

ISSN 1571-3083
Vienna Circle Institute Library
ISBN 978-3-030-97136-6 ISBN 978-3-030-97134-2 (eBook)
https://doi.org/10.1007/978-3-030-97134-2

This Springer imprint is published by the registered company Springer Nature Switzerland AG
The registered company address is: Gewerbestrasse 11, 6330 Cham, Switzerland

Preface

The surfacing of Gödel's part of a planned joint book with Arend Heyting on foundations of mathematics has been one of the pleasant surprises among the papers that Gödel left behind. I have had the good fortune to lead since 2018 a research project dedicated to their study and wish here to thank Tim Lethen for his precious help with the reading of difficult shorthand passages, and Maria Hämeen-Anttila for having prepared the list of Gödel's library loans from the times he worked on the book project.

The Kurt Gödel Papers that this book explores are kept at the Firestone Library of Princeton University. A finding aid with details about their contents is found at the end of the fifth volume of Gödel's *Collected Works*. The papers were divided by their cataloguer John Dawson into archival boxes and within boxes into folders. Folders can have a third division into documents, with a running document numbering system. The papers have been mainly accessed through a microfilm that is publicly available, but also directly in Princeton. References to specific pages of notebooks usually require the use of the reel and frame numbers of the microfilm and that is how the sources are mostly identified in this book. The shorthand manuscript sources are described in detail in Part I, Section 4.1 of this book. These descriptions together with the frame and page numberings in Parts II–III allow the interested reader to identify the source texts with the precision of a notebook page.

The preparation of this book has been financed by the European Research Council Advanced Grant GODELIANA (grant agreement No 787758).

Dedication

This little book was happily recovered from the enormous wealth of materials in Kurt Gödel's papers, in what turned out to be a real detective work. I wish to recollect a decisive circumstance here: The main reason why I was able to learn the Gabelsberger shorthand script in which the pages to follow were written was my exposure to the German language at a proper age, from seven on. Relatively short as this exposure was, it happened at the right time and was quite particular, and the person responsible for it was very special. I recollect with affection Wilhelm Wieczerkowski, my teacher at the German school of Helsinki. To me he appeared just like any other adult – they seemed to be all of one and the same indistinct age – but now I know that he was about thirty at the time and working on his doctoral thesis. The thesis was that bilinguism is not a handicap or a risk that could lead to defects in one's personal development, as the prevailing doctrine had been, but instead an asset. My class was his test object, together with a comparison class in Germany. Here are some of his reflections:

> The thing that makes a difference between instruction given in the German language, and the instruction of a foreign language is [...] in the first place the human encounter between the teacher and his foreign-language pupils. The teacher represents, for the child, and the growing young person, the foreign language circle as well as another way of being. Pupils usually don't make any difference between this foreign way and the person who leads his life accordingly. The deep impressions that come from a personal contact have, without doubt, a long-lasting effect on the attitudes acquired.

After the early career as a school teacher, Wilhelm Wieczerkowski became an esteemed scholar and university professor with highly gifted children as his specialty. I want to dedicate this little volume to his memory, knowing precisely what he would have said had he been confronted with anything of the kind, namely: "Gut Jan, sehr gut!"

CONTENTS

Part I: Gödel's "own book on foundations"

J. von Plato, *Chapters from Gödel's Unfinished Book on Foundational Research in Mathematics*,
Vienna Circle Institute Library 6, https://doi.org/10.1007/978-3-030-97134-2_1

1. A Gödel puzzle

Soon after his astonishing work on incompleteness had been published in 1931, young Kurt Gödel was invited to write together with Arend Heyting a concise overview on foundational research in mathematics. He never delivered his planned part, and that of Heyting got published separately in 1934. It has been often assumed that Gödel simply had not made much progress with the project. The truth, instead, is that he wanted to start by reading practically all of the extant literature on the topics that were allotted to him, namely logicism and the antinomies, and logical calculus and metamathematics. It was an approach that took too much time, and others lost patience at a point where he had practically finished two chapters and planned in detail a third. Thereafter he entertained for a couple of years the idea of "an own book on foundations."

Gödel's great amount of work with the book project did not go lost. It was the basis of his superb, comprehensive knowledge of the logical literature that in turn led him to new discoveries, such as his 1933 work on special cases of the decision problem of predicate logic, and many others that he wrote up but didn't publish. His own views and preferences in foundational matters were informed by alternative ones, works forgotten today that he studied carefully. The thought behind seems to have been, always with an eye for the positive aspects of what others had to present: Could it be that an important or even decisive insight is hidden somewhere in these dusty pages?

For reasons that will be soon explained, the recovery of Gödel's lost chapters on foundational research in mathematics has been quite some puzzle solving in an almost literal sense. To fit his chapters within the bounds set by the publisher, he had to cut out parts of text and write passages in a more compact form. On the other hand, the perfectionist that he was he kept adding materials. Add to this that except for the first part of the first chapter, all others were just first versions. Still, in the end, jumping back and forth, sometimes several pages, and left and right and up and down on notebook page openings, and adding parts from loose sheets, all the bits fit together and coherent, continuous versions of Gödel's chapters miraculously emerged.

The Heyting-Gödel book was meant as a compact introduction to the then current situation in foundational research. Gödel's part was packed with information that would be best read by a fellow logician rather than a newcomer to the field, something that he seems to have realised himself as one letter to the book series editor suggests. Today, when we look at his chapters in a historical perspective and with an interest focused on Gödel himself, the wealth of details just makes it all the more interesting. They show how he perceived the work of others. The two notebooks given in Part III give additional details, many of them about authors completely forgotten today, that did not find their way to the finished chapters.

2. The *Ergebnisse* book project

The Gödel papers contain shorthand versions of an introduction and of two chapters for a small book on the foundations of mathematics, the latter on logicism and the antinomies and on logical calculus. These are found in a notebook that begins with a heavily underlined title "Manuskript für Ergebnisse." There are in addition two plans for one more chapter on metamathematics, found in different places among the papers but clearly identifiable as belonging here.

Gödel had agreed in August 1931 to write together with Arend Heyting a short book on foundational research in mathematics, with publication in Springer's *Ergebnisse der Mathematik und ihrer Grenzgebiete* series. The length of the reports in this series was set as between five and seven printed sheets, i.e., 80 to 112 pages. The Heyting-Gödel book was listed in announcements in 1932 with the title *Mathematische Grundlagenforschung.* Andrei Kolmogorov's famous *Grundbegriffe der Wahrscheinlichkeitsrechnung* of 1933 was another "Heft" in these lists.

Gödel wrote up a reading list for his part, over 130 items, and also more detailed summaries in the notebook *Altes Excerptenheft I 1931–*. A meticulous collector of each and every form of printed or written thing, he kept his library loan requests that can be found among the Gödel microfilm collection. They give us dates for when he read what from the items found in these two notebooks.

Gödel's letters to Heyting, published in volume *V* of his *Collected Works,* show that he had written three quarters of his part by October 1932. In September 1933, he promised to deliver his chapters by the beginning of 1934. At the end of that month, he was to leave for Princeton, where he then stayed until early summer 1934, and wrote from there in January that he would deliver his part by July 1934. Series editor Otto Neugebauer decided at this point to publish Heyting's part separately, with the title:

Mathematische Grundlagenforschung:
Intuitionismus, Beweistheorie

Gödel's correspondence with Hans Hahn, his professor and mentor in Vienna, contains two items that relate to this phase of the book project. Hahn reminds him in a letter of 22 December 1933 about the report and writes again on 2 March 1934 about the "urgent request" for the manuscript on the part of Neugebauer, then that "it seems hopeless that you should finish it during your stay in America."

Gödel's notes give the impression that he, in turn, planned to publish also his part as a separate booklet. Indeed, Heyting's part has two main chapters, about thirty pages each on "intuitionism" and on "axiomatics and proof theory," and some additional material, and Gödel's would have had similarly three chapters. At the end of his introduction, Heyting writes that he does not deal with logicism, problems of pure logic, and metamathematics – exactly the chapters in Gödel's part – and adds that "a separate report is planned on these for this collection." Gödel's title would have been, correspondingly:

Mathematische Grundlagenforschung:
Logizismus und Antinomien, Logikkalkül, Metamathematik

Charles Parsons' introduction to the Gödel-Heyting correspondence gives a discussion of their book project, with details drawn also from letters between the two and Neugebauer. He suggests that Gödel may have been held back "by exceptional meticulousness" which indeed is the case.

Gödel's notes for his chapters give a somewhat bewildering impression. He would write, as was his habit, on the left side of a notebook opening with the right side reserved for additions. A great number of pages have long cancellations at left, blackened to unreadability in what seems quite some work in itself. Often a phrase has just begun, to be interrupted by a "black hole" up to half a page in length, then to continue as if nothing had intervened.

The explanation of the striking cancellations is as follows: The first opening of the *Heft* has the title "Manuskript für Ergebnisse" at right, in a typical mix of longhand and shorthand, and the left side contains the calculations:

$$1\,p = 44 \times 21 \text{ Silben} = 924$$
$$\text{Heyting } 40 \times 19 = 760$$

Gödel had estimated the page length of the *Ergebnisse* book format, by calculating the number of syllables per line times 44 lines, and similarly for Heyting's typewritten manuscript. The number of lines and pages for Heyting's part results in 63 pages for his two chapters and two shorter sections. To remain within his own space limits, Gödel had to cut down the text, which he did by cancelling at left and rewriting in a shorter form at right – or let's say that that would be the typical pattern. At places, he keeps adding. With the two chapters written down, it must have become clear to Gödel that his part, together with that of Heyting, would not fit within the *Ergebnisse* overall limit of 80–112 pages.

The notebook pages show Gödel's own estimates of page length, always written in the lower right corner of the right-hand side of a notebook opening. The numbers for the first chapter on logicism begin with:

$$1, 2, 2\tfrac{3}{4}, 3\tfrac{1}{4}, 4\tfrac{3}{4}, 5\tfrac{7}{8}, 6\tfrac{1}{4}, 7\tfrac{3}{4} \ldots$$

These numbers and similar ones all over in the two chapters on logicism and logical calculus are a clear witness of a basically finished text, with length calculated to one eighth of a page. I say basically, because not all things match, in particular in the chapter on logical calculus, but they could have been fixed at the stage of writing a fair version, something Gödel had time to do only for the first section of the first chapter. I hope this conclusion is warranted by the transcription and translation that I offer. Tim Lethen has provided invaluable help with difficult passages, especially the tiny additions between lines that can be extremely hard to read. Uncertain readings of words have a question mark attached? or ?? in case I did not want to make any guesses. Gödel's own question marks, usually suppressed in shorthand writing, are printed in the German style with a space **?** ahead, and boldface type.

An inspection of the original notebook for the *Ergebnisse* book project, kept at the Firestone Library of Princeton University, shows that the "black holes" in the text are a product of the microfilming process. To obtain a readable reproduction of the text, a high-contrast film has been used, with the described effect on the cancelled passages. They are mostly readable from the original. I have gone through them with an eye on anything that could differ significantly from what Gödel's later uncancelled formulations were.

The introduction and the chapter on logicism and the antinomies, in two sections the first of which comes even in a fair version, are reasonably finished. The chapter on logical calculus is a bit longer, written with much less changes, but unfinished at places. What little exists on the chapter on metamathematics, a list of contents and a couple of introductory pages, is found in a completely different place in the Gödel papers, with a fragment of a letter sketch to Neugebauer as an indication of where it belongs. Remarkably, there is a second start for a chapter–or perhaps a whole little book?–on general metamathematics, not just for arithmetic but for type theory and set theory, in a *Heft* Gödel wrote in Princeton in 1933/34.

Almost all of Gödel's references are indicated by a name and some space. These can be figured out from the context and the two notebooks that list and summarize his sources. I have made some suggestions in clear cases, but have left most of the gaps as they are. All the references are collected together into an extensive bibliography. With its help, those who aim at discussions of specific details should be able to determine what exact passages Gödel has had in mind.

Gödel took up the *Mathematische Grundlagenforschung* initiative and diligently gathered and studied all the pertinent sources, then started to write. Other duties intervened, he was sick for a noticeable time, and he had to make compromises about the presentation. We shall see that the chapter on logicism, followed by his presentation of pure logic, took up all the space that was available, with no possibility to include metamathematics. With his "own book on foundations," instead, he had the space needed, and he had the liberty to present things in the light in which he saw them. For the joint book project, he had first planned a conventional chapter on metamathematics in the style of his 1931 article, but as mentioned, in Princeton he was envisioning a chapter or book on general metamathematics, a treatment that covered type theory and set theory and displayed the implications of incompleteness on mathematics in general, not just for elementary arithmetic. That narrative was to be based on the concept of truth, exactly as his original approach to incompleteness had been, even if that crucial feature was completely suppressed in the final version of his 1931 incompleteness article.

It appears that Gödel entertained the idea of an "own book" about foundations even after his stay in Princeton. His correspondence contains shorthand drafts of letters to Neugebauer, Rudolf Carnap, and Alfred Tarski in which he suggests a new organization of the book. The drafts are somewhat sketchy, with cancellations and incomplete phrases. I shall give these letters here.

Letter sketch to series editor Neugebauer:

> Esteemed Professor!
>
> As you perhaps have heard, I have been ill last summer and fall. The loss of time thus caused as well as other circumstances make it unfortunately impossible for me to work further with the *Ergebnisse* report. If there should be, as I assume, soon a publication, I would like to propose a division of the report among several collaborators. I would like to propose as such R. Carnap (for the chapter on logicism) and A. Tarski for the one on metamathematics, whereas I would myself write a report, some 20–30 pages long, on the present state of the question of freedom from contradiction (newer as well as classical results). Considering the modest scope, it

should be possible to finish the report in a few months, in case the two gentlemen mentioned are prepared to take the task over and in case you are at all prepared to undertake the thing.

I myself had written a great part of the report on logicism, but see now, going through the manuscript, that I had perhaps kept on in too many details of theories.[1] The scope was therefore too strongly limited. I would therefore have to undertake a revision for which, as I saw, time is lacking at present.

with the highest respects, Kurt Gödel

Letter sketch to Carnap:

Dear Carnap,

Many thanks for the complimentary copy of the book as well as the offprints. The rumour has spread here more than once that you would come to Vienna and I hope that it perhaps really comes to that ⟦hard-to-read addition that mentions America⟧.

As you know, I did not deliver any report on foundations for the *Ergebnisse der Mathematik*, as I was preoccupied with other things in the first place. I have made today to Neugebauer the proposal of a division between you, Tarski, and myself, in which you would take over logicism, Tarski metamathematics, and I a report on the present state of the question of freedom from contradiction. I don't know, of course, whether Neugebauer goes for this proposal, but I write anyway so that you have more time to think over whether you want to take care of this. The scope of your part would be about 30 pages.

with best wishes,

Carnap's book has a preface dated May 1934, which means it was published some time in the latter part of 1934. Gödel's help with

[1] There are many changes here, with an approximative reading of what is intended.

the book project is mentioned and one can safely presume that he got the copy upon publication which would place his plans for a joint book on foundations to that time. The letter to Neugebauer mentions his illness in the summer and fall of 1934.

Carnap published in 1935 a review of Heyting's booklet in which he writes that a second booklet by Gödel is in preparation, dedicated to logicism. Carnap kept regular diaries in which further remarks can be found:[2] An entry of 29 June, 1935 contains that "Tarski will write alone the *Ergebnisse* booklet." Another entry, of August 29, 1937, tells that Gödel "has withdrawn his own book about foundations."

From the beginning of 1935 on, Gödel worked for half a year mainly on physics. A notebook otherwise concerned with things such as elasticity theory, thermodynamics, and quantum mechanics, begins with attempts at a formal substitution operator for handling the meaning relation between an expression and an object. There follows a plan of contents for a chapter on antinomies (reel 21, frame 662) and some attempts at handling the epistemological paradoxes formally with the substitution operator.

Gödel's little book contains detailed presentations of his early views on many matters, among them his *completeness theorem for predicate logic* and his views on *set-theoretical relativism*, the latter well before he began his own set-theoretical investigations.

A final question needs to be addressed before we go into the details of Gödel's work with the *Ergebnisse* book. It can be fairly said that both Heyting and Gödel were relative newcomers, for both had in practice just two papers published in August 1931. One can presume that in the case of Gödel, his professor Hahn had already developed great confidence in the young protege's abilities, and may have suggested Gödel to Neugebauer.

As to Gödel and Heyting, they had met each other at the September 1930 Königsberg meeting on the foundations of mathematics. Soon afterwards, Gödel became a household item in logical circles. Heyting, for his part, was the one who had succeeded in presenting intuitionism in an axiomatic form in his path-breaking article of

[2] *Tagebücher 1920-1935,* C. Damböck et al., eds, Meiner Verlag 2021, and *Tagebücher 1936-1970,* Meiner Verlag, in preparation.

1930, a work that intrigued Gödel and led to many of his short, incisive articles of the early 1930s. It is clear against this background that there was a mutual esteem and trust among the two young contributors to the projected *Ergebnisse* volume on the foundations of mathematics. This impression is further confirmed by the recently discovered lecture on intuitionistic logic Gödel gave towards his *Dozentur* in Vienna, on 3 February 1933:

> Brouwer holds as meaningless the claim that each proposition must be either true or false, quite independently of the possibility to establish the one or the other. Brouwer has drawn various conclusions from his claim, in more or less unsystematic ways, and his student Heyting was the first one to attack this question in an axiomatic way. Namely, he put up an axiomatic system of logic, and more precisely, of the narrowest part of logic, the propositional calculus, from which the law of excluded middle cannot be derived.

3. Gödel's reading of the logical literature

After agreeing to participate in the book project, Gödel made a list of literature to study, from titles to mostly short summaries. It is found in a *Heft* with no title (reel 20, frames 499 to 525) and with pages numbered 1–50. There are additional loose unnumbered pages at frames 533–536. The numbered pages 1–50 list over 130 works, with a paper published in 1932 at page 14.

The papers found in the list are directly related to Gödel's chapters. The chapter on formalism was assigned to Heyting, and indeed, none of the formalistic literature is listed but the papers are almost all about logicism, antinomies, and to a lesser extent about logical calculus. This preparatory work seems to be the beginning of Gödel's comprehensive knowledge of practically all of the existing logical literature, an aspect that has not been clear before.

Gödel's notebook with the title *Altes Excerptenheft I* (1931–) is based on the initial list, with more detailed summaries and many more items, also on logical calculus. It appears to be the direct basis for the writing of Gödel's chapters for the *Ergebnisse* report. *Altes*

Excerptenheft I is found in reel 20, frames 236–306. The pages are continuously numbered from 1 to 123, with five more unnumbered pages that make altogether the pages 1–128, seven empty pages at the end, and three small loose notes. There follows a cover page on frame 307 and a separate lined paper page with the names in the *Excerptenheft* listed from Leibniz on page 1 to Sheffer on page 38. Four squared pages follow, a single piece that lists Frege's works (frame 310), and the next opening (311) has the longhand title "Zahlen sind Seiten v. Altes Exc. H **I**" (numbers are pages of the *Altes Excerptenheft I*), and lots of names and page numbers, from 1 to 85, continued on frame 312 with more names and pages within the same range listed.

Many of the items in the two notebooks mentioned are found also in Gödel's library slips. They give a clear idea of when he read what. The library slips have been studied and presented in a tabular form by Maria Hämeen-Anttila. Items with dates have a raised *D* in the text, to indicate that a date is found after the item in the bibliography.

The unnamed notebook contains with most items just the name, title, and place of publication. These are usually drawn over by regular wavy lines, apparently to indicate that the item was cleared in some way or other, as irrelevant or of little importance, or as something on which a summary or remarks are found in the second notebook. Some items, notably those on Ramsey's and Russell's works, have long summaries.

The style of writing of the two notebooks is clearly different from that of the book chapters. In particular, the sentences are typically incomplete, say, the object is clear from the context and is left out, the verb is left to easy guesswork, articles are suppressed, etc. A summary can be a series of brief remarks in the said style. At places, there are formal developments, but most of the formal work is just a recapitulation of the item's content.

The writing of *Altes Excerptenheft I* from page 96 on is markedly different in style. The hand is lighter and more regular, the result very neat in comparison, and the page numbering style clearly different. The last numbered page is 123, but the appearance of the rest is similar to the one from page 96 on. The last page with writing is 128 that begins with: Stellen die für mein Russell Artikel von Wichtigkeit sind (passages that are important for my work on Russell). The conclusion is that the items from page 96 on are from the early 1940's:

Pages 96–109 have summaries of Russell's works with numbered items. Pages 110–114 concern Poincaré's 1906 article in the *Revue de métaphysique et morale,* 24 numbered items.

There follows about one page on Leibniz, and a summary of Frege's *Begriffsschrift* on pages 115–120 in 30 numbered items, usually passages from Frege slightly paraphrased.

There is in reel 32 a list under the title "Alte Literatur Grundlagen" (old literature foundations), with items 1–249. Item 68 has the date 1934, 70 is from 1937, and 90 has the explanation "Vortrag Gentzen 21./IX.1937, soll erscheinen ... Mai 1938," a reference to Gentzen's talk at a conference in Paris. The beginning of this list may be relevant for Gödel's book project.

The two notebooks are given in Part III. There are at places formal details that are not included. Their reading would require a study of the source on which Gödel is making notes. With such a background, the interested reader will likely have little difficulty in going through them directly in the form they are reproduced in the microfilm edition of Gödel's papers.

4. Gödels' manuscript for the Ergebnisse book series

4.1. Description of the sources

The notebook is in reel 24 and begins with inside cover 378L and 378R that has: Manuskr. für Ergebnisse, "für" in shorthand.

Loose pages 379L–381L discuss decidable formulas in predicate logic (379L), then type theory and logicism in an introductory manner (379R–381L). There are two adjacent pages with a red circle, not photographed in the microfilm. I photographed them in Princeton in August 2019. All of these pages can be placed in the text, thanks to special signs Gödel had drawn on them, arrows, coloured circles etc.

Page 381R has a detailed list of contents with the title: A. Logizismus

The title page ("Manuskript für Ergebnisse") has a cancelled text that relates to page 378L. The latter has calculations of the length of Heyting's chapters.

Frame 379 has somewhat oddly at left a page with Van Heijenoort style additions on top, ⟦*iii*⟧ at left, ⟦*iv*⟧ at right, a double bracket style known from *From Frege to Gödel*. The two missing pages with the red circle have similarly the additions ⟦*i*⟧ and ⟦*ii*⟧.

The chapter on logicism has a draft and a *reine* (fair) version. The presence of two versions gives rise to some mildly interesting comparisons, for example, errors in copying, such as the word etwa that becomes etwas, etc. The fair version is written on detached pages the backsides of which contain calculations with partial differential equations. It is a topic Gödel mentions elsewhere as one on which he could give lectures.

Gödel's preserved pages amount to some eighty printed pages. His plans for contents can be gathered from the initial pages and from the correspondence he held with Heyting, all published in volume *IV* of his *Collected Works*. Letters of Heyting to Gödel on 26 July 1932 and of Gödel to Heyting on 4 August 1932 give as planned contents the following:

I Introduction
II Logicism and the antinomies
III Axiomatics and formalism
IV Logical calculus and metamathematics
V Intuitionism
VI Other points of view
VII Relations ⟦between these⟧
VIII Mathematics and natural science

The detailed calculations about length as found on the inside cover page are as follows:

1 p – 44 × 21 syllables = 924
Heyting 40 × 19 = 760   6 N
                                 5 A

Zbl = Heyting −2%
I 29 p H
F $24\frac{1}{2}$ p H

A + N 11 p H
$64\frac{1}{2}$ p H
63 *p*

Here H clearly stands for Heyting, 6 N for six pages on natural science, 5 A similarly for other points of view, Zbl for the page format used by the *Zentralblatt*, I 29 and and F $24\frac{1}{2}$ for Heyting's two chapters on intuitionism and on formalism, and A + N 11 p H indicates that Heyting had finished the short chapters VI and VIII. $64\frac{1}{2}$ p H is the total count of Heyting's part, and 63 *p* the number of pages corrected by two percent. With a maximal number of pages of only 112, Gödel had less than fifty pages at his disposal, a limit difficult to meet.

A letter of Gödel's to Heyting of 15 September 1932 tells that about half of Gödel's part is finished, the section on logicism. Another letter of 16 May 1933 tells that about three quarters were finished by October 1932. A last letter of Heyting's, of 30 September 1933, contains that Gödel had promised to deliver his part by the beginning of 1934. With the idea of a whole book of his own, Gödel started to plan an additional chapter on metamathematics.

4.2. Summary of contents of Gödel's chapters

Gödel's part is divided into three chapters:

There is an introductory chapter of about 3 pages, a chapter on logicism and the antinomies of about 30 pages, and a chapter on logical calculus of about 30 pages – metamathematics not covered. There is a detailed list of contents for the chapter on logicism at frame 381R:

1. Logicism and the antinomies

p $20\frac{3}{4}$–$23\frac{3}{4}$ not interesting

A. Logicism

1. The claim of logicism – what is logic – Russell earlier – now – Wittgenstein – Hahn
2. Necessity of a symbolism for the execution of the program – *calculus ratiocinator*
   Formal system – the first such calculus Peano shows that the whole of mathematics follows from a few assumptions
3. Execution of logicism Frege Basic concepts, axioms, strict

formalization

4. Definitions in various forms
5. Derivation of mathematics
   a. Natural numbers Dedekind – Frege Russell – Peano, Cantor definition of finite, infinite, Tarski
   b. Negative numbers, fractions (Tannery) Peano Russell
   c. Irrational numbers
   d. Measure theory
6. The antinomies and general method for producing them
7. Russell, different ways for the possibility to overcome them Execution of zig zag in Frege, Russell *Principles*, Behmann – about limit in the axiomatics of set theory, no class theory
8. Poincaré and non-predicative definition, discussion with Zermelo etc
10. Presentation of the Russellian theory as well as the axiom of reducibility, descriptive functions, typical ambiguity
10a. 2. edition Russell
11. Formalization by Lezniewski
12. Chwistek
13. Improvement by Ramsey
14. Carnap, Langford

The effect of Gödel's thorough acquaintance with the logicist literature is visible in his 1933 Vienna lectures on the foundations of arithmetic, as well as in his article on Russell's mathematical logic.

2. Logical calculus

The contents of the rather complex chapter on logical calculus, with frame numbers, are as follows:

A paragraph about the concept of formalization, 420
Tarski's general theory of consequence, 2 pages, 420–421
Formal systems, 1 page, 421-422
Propositional logic: various systems, completeness, independence, 4 pages, 422–426
Many-valued logic, 1.5 pages, 426–427
Subsystems of propositional logic: positive logic, Hertz systems, 0.5 pages, 428
Quantified propositional logic, 1 page, 429

Strict implication, modal logic, 2 pages, 430–431
Predicate logic: the language, Skolem-Löwenheim, completeness, first-order theories, denumerable models of set theory, finitary consistency proofs, 8 pages, 432–440
Decidability: the general problem, special decidable cases and theories, 4.5 pages, 440-444
Algebraic logic, 8 pages, 445–452

Logical calculi were extensively covered in Hahn's seminar on mathematical logic of 1931–32. Gödel was a very active participant there and spoke on eight occasions, among others about many-valued logic and consistency proofs, and the seminar clearly had a strong influence on Gödel's chapter on logical calculus. He also presented his results on intuitionistic logic at the seminar, but left that topic out from the Heyting-Gödel volume. His 1933 paper on the decision problem of predicate logic was inspired by his reading of Behmann and others on special cases of the decision problem, a topic presented in detail in the chapter on logical calculus.

3. Metamathematics

Gödel got just to the beginnings with the writing of his chapter on metamathematics. The above two long chapters had already more than exhausted the space at his disposal. It is not clear how he thought he could accommodate all the materials he needed to cover, a problem solved by the thought of writing a whole book of his own about foundations. What little is found on metamathematics is described as follows:

1 §. *Sketch for a chapter on metamathematics.* Reel 47 of the Gödel microfilm collection contains varied materials, among them a concentration of letter drafts and related materials. Along with what is preserved of a letter draft to Neugebauer, there is a plan of contents and a beginning for what clearly would have been Gödel's chapter on metamathematics for the joint book with Heyting (frames 529–530). The list of contents is detailed, with 11 sections, but the text itself is very short.

2 §. *Sketch of a book on general metamathematics?* The Gödel papers contain a notebook with the title "Amerika 1933/34" that begins in English with the subtitle "Metamathematical Notions" (reel 19,

frame 755), with more than forty pages following in a mixture of German and English, some English sentences with occasional German words written in Gabelsberger. These notes are clearly made for Gödel's well-known lectures on his incompleteness theorems at Princeton, from February to May 1934. There follow four pages on propositional logic with the name Wajsberg written on top. The text now hits the end of the backward direction of the notebook that begins with the title "Arithmetik Amerika 1933/34" (frame 783). The topic is again incompleteness, mixed with several pages of geometric diagrams and partial differential equations, until a plan of contents for a book or other longer treatise on frame 792R. The text proper is only five pages long. Then follow three pages of formulas about incompleteness, and the notebook ends.

Part II: Own book (foundations)

J. von Plato, *Chapters from Gödel's Unfinished Book on Foundational Research in Mathematics*, Vienna Circle Institute Library 6, https://doi.org/10.1007/978-3-030-97134-2_2

24-384

⟦1. Introduction⟧

One will want to request from a foundation for mathematics, above all, answers to the following three questions:

1. What is the epistemological nature of mathematical theorems? Two conceptions remain in consideration today, namely that they are analytic, and that they arise from a form of pure intuition ⟦reine Anschauung⟧.

2. What are the mathematical concepts, propositions, and proof methods that one can assume to be secured on the basis of the answer to 1, i.e., what are the tautologies not further reducible and what the propositions given in intuition, respectively?

3. How can one derive the usual mathematics from these basic laws, and how far does it lead?

The first question is the one that occurred historically first. The two contrasting conceptions are connected to the names of Leibniz and Kant. They are embodied today in logicism on the one side and in intuitionism and formalism on the other. The contrasts are as rough as ever and one can see progress at most in the strive to more precise formulations of the standpoints. As concerns the second question (in the second half of the 19th century), it is interesting to note that the answer is largely independent of the first one. Thus, Weyl in his *Kontinuum* on the one hand (and then Chwistek), and Russell in the second edition of the *Principia* on the other hand came, starting from two different conceptions of the nature of mathematical theorems, to closely equivalent basic laws.

One can say quite generally that the three main directions

24-385

converge more and more in their answers to question 2. Formalism and intuitionism agreed on it all along and logicism that assumed earlier certain basic laws that were rejected by the other directions (e.g., the reducibility axiom), has given up in part in the past years (*Principia Mathematica*) and shows in its leading representatives a strong constructivistic tendency (cf. *Königsberg*, p. 104, Hahn, p. 138.)[1]

The third of the problems put up above can be answered in two ways, by deriving from the basic laws either the theorems of mathematics themselves, or just their freedom from contradiction (mutual compatibility). Intuitionism does the first but in so doing has to renounce essential parts of mathematics, as they are not derivable from its basic laws. The formalistic school attempted the second and hoped thereby to justify the whole of classical mathematics, though admittedly only in the sense of freedom from contradiction.

Logicism had initially in its program the deduction of the theorems of classical mathematics itself (not just its freedom from contradiction). This becomes impossible under the constructivistic tendency mentioned above, so one tends to be satisfied even here with freedom from contradiction. According to Hahn (*Königsberg*), for example, analysis is to be founded as a hypothetico-deductive system. It is not the theorems of analysis themselves, but just what follows from the axioms that can be founded logically, and a proof of freedom from contradiction for the latter is therefore required. There follows a notable convergence with formalism.

Carnap presented the view (*Königsberg*) that there should be no essential difference between the two described ways of solving problem 3,

24-386

because a proof of freedom from contradiction must always offer an interpretation of the basic concepts to which the axioms to be shown free from contradiction apply.

Whereas questions 2 in so doing ⟦should be 1⟧ and 3 ⟦should be 2⟧ are of a philosophical nature, problem 3 is a purely mathematical one, as is shown by this being the problem above others with which results generally recognized were achieved. One can express the problem in a generalized form as something like: Which mathem-

[1] ⟦The page reference is to Carnap's article in the proceedings of the September 1930 Königsberg meeting, in *Erkenntnis*, vol. 2, 1931⟧

atical theorems follow from what basic laws ? It was put down by *Menger* in this form, as the sole meaningful problem of mathematical foundational research:

"Implicational point of view"[2]

A great number of other investigations are counted among mathematical foundational research, beyond the problems mentioned above. The latter have been in part a point of departure for these investigations that have actually nothing to do with a "foundation," but present instead a discipline within mathematics. They make up, nevertheless, the greater part of the literature, interesting for the mathematician. Herein belong investigations about the independence of single basic laws, further the question whether and how the propositions of a given domain can be decided from some axioms, as well as to deliver a procedure by which such a decision can be carried through. Further, a theory that studies the general properties of formal systems and systems of axioms (cf.   ) (axiomatics), and several others. All of these investigations are brought together, as far as possible, in Section ⟦Logical calculus⟧.

---

[2] ⟦Changed from: implicationism⟧

24-387

⟦2.⟧ Logicism

Under logicism is understood the conception, going back to Leibniz and founded in detail by Peano, Frege, and Russell, by which mathematics is without residue reducible to logic, i.e., that the mathematical concepts are definable from logical ones and the mathematical theorems provable from the axioms of logic (Carnap). What is to be understood by logical concepts and propositions in this? Further, how do these distinguish themselves from other concepts and propositions? This question remained unanswered by logicism itself for a long time. It remained content to list what are known as "logical constants," or, not, all, there exists, etc, as well as logical axioms. A certain kind of intuition was assumed as a criterion for the truth of a proposition of logic (cf. Russell). One held, though, as justified even basic laws that were not intuitively obvious, by the fact that they led to the desired results that agreed with intuition (cf. Russell's axiom of ordering, *Revue de métaphysique et morale* 1906, p. 630). (Russell puts up explicitly the theory that just as hypotheses in physics cannot be directly tested, but are instead justified by their leading to results that agree with intuition.) *Frege* saw, instead, the logical basic laws as lawlike regularities that obtain in the realm of ideas and the justification of which is not something one can ask for. A satisfactory conception was put forward only in recent times under the influence of Wittgenstein (Russell , Hahn). According to this conception, the characteristic of logical constants is that they are not signs for something that is present in the world, but they serve just to express the forms of sentences.

1

24-388

They are needed only because of the complicated way in which language associates its symbols to reality. This association is not one-to-one, so that it is possible to express the same state of affairs in different ways (e.g., by p and $\sim\sim p$), and the theorems of logic consist properly in determining that two combinations of symbols become associated to the same state of affairs (e.g., $p \equiv \sim\sim p$). This tautological character is the criterion for the theorems of logic. The thesis of logicism becomes thereby the claim that mathematical theorems are tautologies. (It follows from the thesis that they must hold in each possible world). This states essentially more than the original Russellian formulation of mathematical theorems, in which there occur only logical constants, because even empirical facts can be expressed in such theorems (as for example in the proposition that there exist at least ten things) (Ramsey).

Wittgenstein and Ramsey tried to make the concept of a tautology more precise in the following way: They assumed there would be certain simplest propositions (atoms, say: This is red.). The truth values true and false can be distributed in different ways over these atomic propositions (there are with n atomic propositions 2^n possible distributions). Each meaningful proposition asserts, now, that the factual distribution at hand belongs to a specific subset of the set that comprises these 2^n elements, something that was expressed as follows: Each proposition is a truth function of the atomic propositions. In case T comprises all the possible distributions, the proposition is a tautology. If T is empty, it is a contradiction (say p and not p).

To prove the thesis of logicism as initially formulated, it is absolutely necessary, as already remarked by Frege ( ), to replace the imprecise and equivocal word-language by a precise formula-language. Is it possible to represent the logical basic laws in a precise way through formulas, and to state exactly the logical rules of inference by which one derives further formulas from these, thereby to avoid that extra-logical elements creep in unnoticed during deduction? Rules of inference are purely formal here, i.e., they refer only to the external shape of formulas, not their meaning, so that inference turns into a calculation.

The idea of a formal system arose in this way, already anticipated by Leibniz in his *calculus ratiocinator* and then brought to full clarity by the Hilbertian school. The founders of the algebra of logic (Boole, Peirce, Schröder) provided valuable preliminary works for this formula language (concept notation, pasigraphy), but it was only Peano who gave it the flexibility needed, to make possible the expression of quite complicated mathematical propositions. He and his collaborators

2

24-389

presented in (*Formulaire*) numerous fields of mathematics (number theory, arithmetic, analysis, geometry) in a purely symbolic form. The logicist program was, nevertheless, carried through only to the point that the theorems and concepts of the field of mathematics in question were to be reduced back to a minimum of undefined "concepts" and unproved "propositions" (compare, for example, the well-known Peano axioms for arithmetic), without striving at a reduction of these ultimate elements to logic. Moreover, as concerns rigour in deduction, the system of Peano in no way satisfies the requirements of today (cf. Frege's letter to Peano, *Rivista di Matematica* VI, p. 53). The rules of inference are not formulated in a sufficiently precise way to render possible a purely mechanical application, and the way in which new signs are introduced through definitions is, in particular, arbitrary, so that one should in this connection often speak of an axiom.

The execution of the logicist program was, admittedly, not the main aim. Peano wanted rather

$2\frac{3}{4}$

24-390

to create, in the first place, an encyclopaedia of proved mathematical theorems (together with detailed historical notes), something that became possible in a relatively small space because of the abbreviating effect, in about the relation 1:10, of the formula language. Most theorems were therefore introduced without proof. Peano hoped further to achieve, through the introduction of an appropriate symbolism, a real advancement in the ease of mathematical inference, say like the one achieved by the decimal system for calculation with numbers.

Frege provided in [⟦*Grundgesetze I*⟧] the first carrying through of the logicist program. He achieved in the formulation of the basic laws laid as a foundation a precision that satisfies even the most modern requirements and that was reached later only by the formalist school and the Polish metamathematicians. The logical basic concepts that he laid down as the undefined ones of his deductions are the following:

1. "Objects,"

$3\frac{1}{4}$

24-391

i.e., arbitrary objects of thought. This concept has no specific sign in the system but appears only through variables $a, b, c \ldots$ that designate arbitrary objects.

2. "Function." By this is understood an arbitrary expression, in which one or more object variables a, b occur, that designates a specific object as soon as the variables are substituted by names for specific objects (e.g., "father of a," $sin\, a, a + b$). The expression that arises through the substitution of specific objects is, then, according to Frege, a composite name for the object, called the value of the function for this argument k. Frege even conceives of propositions as composite names, namely, the true propositions as names for the True, and the false ones as names for the False, respectively, by which both "truth values" are conceived of as two specific objects (cf.   ).

If a function is so constituted that a proposition arises when specific objects are substituted for variables (e.g., $x > y$), it is called a propositional function. Its values are in this case truth values.

$3\frac{3}{4}$

24-392

If functions of this kind contain a variable, they are called "properties," and if they contain k variables, k-place relations.

Frege does not count functions among the objects, with the justification that they are incomplete (amenable of completion), namely through the substitution of a name. A very clear distinction is made between functions the arguments of which are objects and those the arguments of which are again functions (see below) (⟦*Grundgesetze I*⟧, p. 3, p. 36).

Even the concept "function" has no special sign in the system but occurs only through a sort of variables $f, g \ldots$ and M, N, respectively, that denote arbitrary functions of the first and second order, respectively.

There occur as basic concepts, besides concepts 1 and 2, a series of special functions each of which has a particular sign, namely:

1. $-x$ x is the True

2. $\sim x$ x is not the True[3]

3. $x \supset y$ y follows from x, i.e., it is not the case that x is the True and y not the True

Further the theory that one can build up with just the two concepts 2 and 3 (the propositional calculus, cf.).

4. $x = y$ x is identical to y (i.e., x, y are names for the same object)

Further two functions of second order (5., 6.):

5. $(x)f(x)$ $f(x)$ is for each object the True

The proposition that there exists at least one x for which $f(x)$ is true can be easily expressed with the help of this sign (all-sign), defined in the form $\sim (x) \sim f(x)$ (abbreviated after Russell as $(\exists x)f(x)$).

6. $\acute{x} f(x)$ the course-of-values of the function $f(x)$

The meaning of course-of-values comes out from the axiom that stands below. It states that two functions f and g have the same course-of-values if and only if

$4\frac{3}{4}$

24-393

$f(x) = g(x)$ for all objects x

---

[3] ⟦*Altes Excerptenheft*, pp. 27–30, has a summary of Frege's *Grundgesetze I* with: $\top x$ x is not the True⟧

The course-of-values corresponds therefore in the case of concepts to what is usually called the "scope" of a concept, or the "set of objects that fall under a concept." Frege counted the course-of-values, contrary to the function itself, among the objects. That had ominous consequences for what are called antinomies.

7. $\imath x$

$\imath x$ means: If x is the scope of a concept under which there falls exactly one object, that object. In all other cases, $\imath x$ means the object x itself.

One obtains then, through the combination of the concepts 1–7 and the variables, all possible meaningful expressions of the theory (compound names in the Fregean terminology). Frege specifies with the most painstaking care how this combination has to be done, to obtain meaningful expressions (correctly built names) (*Grundgesetze I*, p. 45). In particular, the argument places of functions of the first order can be filled in in a meaningful expression only by objects (and object variables, respectively), and in ones of order two only by functions (function variables) of order two ⟦one⟧. (Cf. Russell's type rule, p.   ).

A variable in an expression is called bound if an all-sign or a course-of-values sign relates to it, otherwise free. The scope of an all-sign or a course-of-values sign is the subexpression that begins with this sign, uniquely determined as one can show. A simpler name can be introduced *per* definition for compound names of objects (functions). Therefore definitions don't accomplish anything, but just rename what is already at hand, and each definition has therefore the form:

$$n = \ldots \qquad f(x) = \ldots \qquad g(xy)\ldots$$

Here n, f, g are the newly introduced signs, and the right side has names correctly built from the basic signs and of previously defined signs.

Frege expresses through $\vdash A$ the assertion that a compound name A denotes the True ($\vdash$ is called the assertion sign).

If there occur free variables in A, then $\vdash A$ means that the expression A always denotes the True, irrespective of the specific objects (functions) one substitutes for the free variables that occur in it.

The following eight assertions are put at the top as axioms ——

The definitional equations that were discussed above count in addition as points of departure for deductions. The formal rules by which one goes on to derive from the initial formulas are given with the utmost care. The most important of these are:

1. The rule of substitution: A free object variable can be substituted, in an axiom or an assertion already proved,

$5\frac{7}{8}$

24-394

by an arbitrary correctly built name for an object (and an analogous rule for function variables).

2. One can infer $[\![\vdash]\!]B$ from $\vdash A \supset B$ and $\vdash A$, with A, B arbitrary meaningful expressions.

Frege assumed, beyond 1 and 2, a number of further rules of inference, among them:

$A \supset C$ can be concluded from $A \supset B$ and $B \supset C$

Further, $B \supset . A \supset C$ can be concluded from $A \supset . B \supset C$

Further the rule that concerns the all-sign, p. .

Frege fulfils, in a close to perfect way, all the details one poses today on a formal system. It is his particular merit to clearly distinguish, as the first one, between the assertions and concepts that can be expressed in the logistic symbols proper, and those considerations that are formulated in ordinary word-language and that deal with the symbols of the system (Hilbert's

$6\frac{1}{4}$

24-395L

metamathematics). To the latter belong, for example, the definition of "meaningful" by which the grammar of the symbolic language is specified, and the formulation of "the rules of inference." This distinction was in part blurred by his successors (B. Russell, Whitehead), to the *detriment* of clarity.

Frege used his system in the first place to develop a purely logical theory of natural numbers (later independently discovered by B. Russell).

Cantor had defined the cardinal number of a set as the general concept ⟦*Allgemeinbegriff*⟧ it has in common with all equivalent sets (*Annalen* 21). This is what is known as a definition by abstraction, the nature of which was as well first analyzed by Frege. It consists in associating to each thing x of a domain B in which there is defined a symmetric, reflexive, and transitive relation R (in our case equivalence), a new thing $\varphi\, x$ (in our case, the cardinal number of x) in such a way that $\varphi\, x = \varphi\, y$ if and only if R obtains between x and y (other examples from mathematics: vector direction, shape).

This definition is seemingly not a mere giving of names, but it creates new things, the $\varphi\, x$.

24-380R[4]

Still, Frege, and independently of him Russell [ ], show that it can be subordinated to the above scheme, in a nominal definition, if one understands by $\varphi(x)$ the set of things from B that stand in the relation R to x. One arrives in this way at the:

Definition. A number is a set that consists of all the sets that are equivalent to a set.

But the essentially new achievement of Frege's, in comparison to Cantor, consisted in a definition of finite numbers and the logical derivation of complete induction that it made possible. A property (set) is said to be hereditary if, whenever it belongs to a number x, it always follows that it belongs to the number $x + 1$. A number is said to be finite if it has each hereditary property that belongs to 0, i.e., if it belongs to each hereditary set that contains 0. The theorem of complete induction follows immediately from this, and the rest of the Peano axioms follow equally easily (in the system of Frege). Frege proved especially that each number has a successor

---

[4] ⟦Page ripped off the notebook. A heavily drawn sign on page **395L** and this page connects them. ⟧

(i.e., that there exist infinitely many numbers), by showing that the set of numbers from 0 to n has the number $n+1$.

Objections to the Frege-Russell definition of natural numbers were raised above all by Poincaré. He accused it of circularity, as the natural numbers are already presupposed in the basic concepts of logic. One speaks about relations with two terms, products of two propositions, etc, to which Couturat justly objected that this apparent circularity should arise from the imperfection of the word-language, something that disappears with a symbolic representation, and further, that one should not need the concept of number to think of two objects being in a relation (*Revue de métaphysique* 1906, p. 221).

Poincaré's objections to the logical derivation of complete induction (cf.) have been widely recognised. He took it to be a synthetic judgment *a priori* and saw in it the principle by which mathematics gets distinguished from logic (cf.).

Dedekind tried in ⟦1888⟧, almost at the same time as Frege ⟦cancelled: and independently of him⟧, to build up the theory of natural numbers purely logically,[5] with the derivation of complete induction almost identical to Frege's. ⟦Cancelled: He lays here as a basis the concept of a set (system) and a mapping.⟧ He assumed a set Σ and a mapping φ ⟦and?⟧ simply required: If there exists an element $a \, \varepsilon \, \Sigma$ such that the system Σ is mapped by φ in a one-to-one way on $\Sigma - \{a\}$ and if, further, no proper subsystem of Σ ⟦added in margin: that contains 0⟧ is mapped to itself by φ. From this, the theorem of complete induction immediately follows. He defines the natural numbers in an imprecise way as that which results from an infinite system when one disregards the special nature of the elements. Beyond this weak point, Dedekind's development satisfies high requirements of rigour. Dedekind was, especially, the first one to recognize that the definition through complete induction, as used first by Grassmann and later by Peano for the introduction of the arithmetic operations (e.g., multiplication through the two requirements $a \cdot 0 = 0, a \cdot (b+1) = ab + a$),[6] is in need of a justification, i.e.,

---

[5] ⟦The clear version reaches this point. What follows has lots of changes and seems incoherent at places.⟧

[6] ⟦Changed from: addition through the two requirements $a + 0 = a, a + (b +$

of a proof that there exists one and only one function that satisfies the recursion axioms. Cf. on this [Dedekind], Landau ⟦(1930)⟧, ??ayr, v. Neumann *Annalen 99* in which the same problem

24-381L

was resolved for transfinite induction. Dedekind defines, in contrast to Frege, the finite sets as those that are not similar to any of their proper parts, ⟦added: (On the relation of the two, cf.)⟧ and tried to prove the existence of infinite sets by the following: There must exist to each thing a concept, by which the set of all things is mapped into a proper subset of the concepts (this goes back to Bolzano §). [Cf. p.]

Thus, the arithmetic of natural numbers was reduced by Frege, Dedekind, and Russell back to pure logic, and since one had already reduced the higher kinds of numbers (negative, fractional, irrational numbers) purely logically to the natural numbers, (cf. about that), it seemed that the logistic program had thereby been carried through. – It then happened by the turn of the century that these set-theoretical ways of inference, the ones that made possible the above derivations, led under unlimited application to contradictions, and that these contradictions could even be deduced from the above axioms of logic of Frege.

$1) = (a + b) + 1$⟧

⟦3.⟧ Antinomies[7]

The historically last of these set-theoretical antinomies is the inconsistency of the greatest ordinal number:

1. The set of all ordinal numbers ordered by the relation of "greater" is well-ordered, therefore must itself have an ordinal number that must be greater than all ordinal numbers, then even greater than itself. Burali-Forti.

A great number of antinomies has been constructed on purpose ever since, to discover the cause of the phenomenon:

2. The cardinal number of the set of all sets A clearly has to be the greatest cardinal number, because each set is a subset of A. Cantor has, on the other hand, proved that there exists for each cardinal number α a greater one, namely 2^{α} (Russell *Principia*).

24-395R

Russell, in analyzing this contradiction, i.e., in applying the Cantorian proof for $2^{\alpha} > \alpha$ to the set of all things, finds his well-known contradiction:

3. The set of all sets that do not contain themselves as an element. One can show that it can neither contain nor not contain itself.[8]

To show that this antinomy does not depend on the concept of set, Russell gives it in the following form [*P.L.*]:[9]

4. A property is predicative if it applies to itself, otherwise impredicative (e.g., abstract is ⟦predicative⟧, concrete is ⟦impredicative⟧).[10] One can show of the property "impredicative" that it is neither predicative nor impredicative.

5. The relation T that holds between two relations R and S if and

---

[7] ⟦Continuation of page **395L**. Gödel gives on page 453 a list of antinomies 1–11 that matches the numbering below. The ones that don't get a separate number here are 7. The heterological paradox. 8. A paradox about the set of all propositions. 9. Smallest number definable in less than a hundred words. 11. Richard's paradox.⟧

[8] ⟦Text continues on the next page, with the indication: (principal)⟧

[9] ⟦Russell (1906c)⟧

[10] ⟦In 4, autological and heterological have been changed into predicative and impredicative. The example was first: abstract is autological, concrete is heterological.⟧

only if R does not have the relation R to S leads to a contradiction similar to 3.

The antinomies mentioned so far have in common that they operate with purely logical (arithmetical) concepts. They surface therefore within the formal systems of mathematics (unless special care? is applied), for example, within the Fregean system (with the exception of 4 that by the Fregean distinction into concept and object does not occur).

A second kind of antinomies operates with empirical (linguistic) concepts, "*A* asserts that," "words of the German language," "sentence of a certain language," and similar. They are therefore called epistemological. (The distinction goes back to Peano's *Rivista*. They were first clearly formulated by Ramsey.)

6. To this group belongs the liar, when a person *A* says: The sentence I now state is false. This sentence can be neither true nor false (cf. Hilbert-Ackermann, p.   for a logical formulation of this antinomy). (A work about this with Finsler.) This is nothing else than the "Cretan inference" that, however, in the form it has been handed down in which a Cretan says "all Cretans lie" leads to no contradiction but just to the consequence that some Cretans lie and some don't lie.

$7\frac{3}{4}$

24-396L

⟦9.⟧ There are in the German or any other language only finitely many sentences that consist of at most a hundred words. There must then exist a natural number and even "a smallest number among the numbers that cannot be defined with at most a hundred German words." This number, however, is defined by less than a hundred words, by precisely the words that stand under the quotation marks (Dixon   ). This antinomy is presented symbolically in Hilbert-Ackermann (1928, p. 95) in which the ambiguous concept "definable by a hundred words of the German language" is replaced by the sharper "number defined in the 20th century." The cause operates manifestly within the finite, and Russell makes therefore use of it in ( ) against the view represented by Poincaré that the actually infinite should be responsible for the antinomy.

10. The contradiction of the smallest ordinal number that cannot be defined in the German language is of a character similar to 2. (There must be such a number because of the non-denumerability of the set of ordinal numbers and the denumerability of the sentences of the German language.) It is also similar to the antinomy of Richard ⟦11⟧ that arises when one thinks of all the

$8\frac{3}{4}$

24-397L

definitions of infinite decimal expansions as lexicographically ordered, and then forms by the diagonal procedure an α that is different from all the decimal expansions definable in the German language. This decimal expansion α is, on the other hand, defined by the following German words:

> α ist der nach dem Diagonalverfahren aus der lexicographischen Folge der in deutscher Sprache definierbaren Dezimalbrüche gebildete Dezimalbruch.[11]

⟦7.⟧ From Grelling ( ) comes an epistemological analogue to 4: A property word shall be called autological or heterological when the property expressed by the word applies to it and does not apply to it, respectively. Example short and long ⟦kurz, lang⟧, respectively. Heterological is, then, neither autological nor heterological.

⟦8.⟧ Russell (*Principles*, p.) gives an epistemological analogue to 2: One can associate in a one-to-one way to each set $\mathfrak{m}$ of sentences the sentence by which all the sentences in the set are true. Therefore there exist *at least as many* sentences as sets of sentences. On the other hand, there must exist by Cantor *more* sets of sentences than sentences. Here belong further the mentioned antinomies of finite description 9–11.

$9\frac{1}{4}$

---

[11] ⟦α is the decimal expansion built by the diagonal procedure from the lexicographical sequence of decimal expansions definable in the German language.⟧

24-398L

Russell indicates in   the general circumstance that lies at the basis of all the antinomies. This circumstance consists in there (apparently) being certain concepts Φ with the property that each set u, all the elements of which have the property Φ, gives rise to the possibility to define a thing $f(u)$ that has the property Φ but that lies outside of u. The set φ of all things with the property Φ then leads to a contradiction, because $f(\varphi)$ cannot belong nor not belong to φ. (Φ is called self-reproductive.)

Φ is in 1 the set of ordinal numbers, in 2 that of the cardinal numbers, in 3, 4 the set of all sets and of all concepts, respectively, in 6 the set of all sentences, in 7–9 the set of all definitions for a certain kind of object $f(a)$ in the German language. Meaning in 3 is, for example: The set of sets from u that don't contain themselves, of which one can show that it does not lie in u. A more precise analysis reveals that there occurs in fact in 6 the set of all sentences. Namely, its logically correct formulation is as follows: "There exists a false sentence that I assert now." If one says instead: There exists a false sentence from the set $\mathfrak{m}$ that I assert now, then one has the above process $f(m)$ that associates to each set of sentences a sentence that cannot lie in it.

10

24-399L

A somewhat different scheme for antinomies is found in Grelling-Nelson.

It was clear after the discovery of the antinomies that one had to restrict in some way the rules of "naive logic," the ones that one had so far applied with no caution within mathematics (especially in set theory) and that Frege had formulated in his system in a precise way. The latter does not correspond quite precisely to the naive standpoint, because antinomy № 4 is already excluded through the distinction of properties of different orders. For a formalization that reproduces precisely the naive standpoint, cf. Hilbert-Ackermann (p.   ).

One can distinguish the pertinent approaches into conservative and radical ones by how much of the naive logic is maintained. Two of the most conservative ones were attempted in the past few years (Behmann, Church), but no conclusive judgment is yet possible.

Behmann believes that the paradoxes arose by a lack of attention to the highest rule for the introduction of new signs (abbreviations) through definitions, namely that an expression in which there occurs an abbreviation has sense only if one can eliminate this abbreviation (i.e., can write it also without the use of such).

The property of not belonging to oneself (compare antinomy number 4) may be correct and one could also introduce an abbreviation F (impredicative) for it through the definition $F(\varphi) \equiv \varphi(\varphi)$. The proposition $F(F)$ could not, in contrast, be written without the use of the abbreviation F and would therefore have to be considered meaningless. For if one substitutes the first F through the definition, one obtains the same expression again. The substitution of the second F would, on the other hand, be impossible at the outset. (To this would be objected that if the substitution of a variable at an argument place by a specific function is declared impossible, even most mathematical propositions would become meaningless.) In any case, the limitless use of naive logic can lead to contradictions even without the use of abbreviations

$11\frac{1}{4}$

24-400L

(cf.   ), and Behmann finds himself therefore compelled to limit these rules (e.g., *modus Barbara*). He tries, however, to derive this modification even from the principle of abbreviation. Namely, as a consequence of this principle, the all-operation has a limited meaning, because one would be allowed to substitute only such symbols for the all-variable for which the result of substitution can be written without abbreviations. This doesn't have to be the case for certain singular values of the variables. The precise rules of the formal system to which one arrives (called ultra-finite logic by Behmann), starting from these insights, is still lacking.

Church (   ) offers a formal development of a theory that lies close to Behmann's (even if not in its starting points). One can build also in his system the concept F (impredicable), but one cannot prove that $F(F)$ is either true or false. This just means that $F(F)$ is meaningless. The resulting great difficulty can be fathomed from the fact that 37 axioms had to be put on top and the author himself could not decide in his first publication (   ) whether the antinomy of Burali-Forti arises in the system.

Here belongs even the attempt at a solution undertaken by Frege himself. It consisted in substituting the above axiom   by the following one (naturally symbolically formulated in Frege): The scopes of two concepts f, g are the same if each object (with the possible exception of these two scopes themselves) that falls under one of them falls also under the other one. The contradiction is in fact thereby avoided. Whether this is the case for the others was not investigated.

$12\frac{1}{4}$

24-401L

The three attempts at a solution discussed so far have this in common, namely that the construction of the concepts (and sets, respectively) that lead to a contradiction remain possible, and just some propositions (inferences) about them become excluded. Russell, on the contrary, represents the opinion that there could not exist things like the set of all sets, of all ordinal numbers, cardinal numbers, and similar. He develops this conception in *P. L. M.* and *Rev. met.* in the first place for paradoxes of the first kind (the semantic ones were at the time still largely unknown). The contradictions have shown that there cannot exist for each conceptual property (propositional function) a set in which all things with this property would be collected (for example, to be an ordinal number, cardinal number, set), or in other words (because one can substitute in the paradoxes the set by the propositional function itself, cf. An⟦tinomy?⟧ 4), that one cannot consider each propositional function as separable from its argument. One could now either search for a criterion on propositional functions for which this case occurs, or one can declare that sets and propositional functions don't exist as objects at all, and that propositions about them are to be considered an improper mode of speech about their values and arguments (no class theory). For example, the sentence that there exists a concept under which *a* falls would mean that there exists a proposition that becomes a true one when a component therein is substituted by *a*.

The first, conservative way is the one later taken by the axiomatics of set theory. Its axioms are in fact nothing else but stipulations of when there exist sets that belong to propositional functions.

13

24-402L

Russell himself does not follow this way further. He makes do by sketching roughly two criteria. One of them (cf. *Principia*, p.) was never studied any closer, and the other one states that for a concept to determine a set, it should not be too big, or more precisely stated ( ), it should be mapped in a one-to-one way to a proper section of the series of ordinal numbers.

Russell's idea finds its precise execution in von Neumann's axiomatics of set theory, one axiom of which, transferred to ordinary language, states that a set exists if and only if it is not equivalent to the universal set. It is especially interesting that, as von Neumann shows, the axiom of choice follows from this axiom.

Russell's own investigations about the paradoxes move in the direction of the no class theory and in the end to the putting up of his "type theory." (Cf. for the way, not entirely transparent, that leads from the former to the latter.) The starting point of type theory is that the objects considered in science fall into basically different kinds (types). The simplest, ones that cannot be analyzed further, are called individuals, the ones next in complicatedness are the classes of individuals and relations between individuals. There exist further classes of classes of individuals, relations between classes of individuals, and so on *ad infinitum*. Type theory states further that a property that one can meaningfully ascribe to a thing of a determinate type cannot be ascribed to a thing of another type, i.e., if one, say, substitutes in a proposition the name of an individual by one of a class, no proposition results, but a meaningless combination of signs.

14

24-403

If one designates as the "range of values" of a variable the set of those things that one can meaningfully substitute for it, it follows from what was said that a variable can have as a range of values just one specific type, and therefore also the operations all and there exists can refer to only one specific type. This rule and a respective formal system that is built on it as a foundation is fully sufficient for avoiding antinomies of the first kind. Antinomy $N^{\underline{o}}$ ⟦1⟧, for example, is resolved through the impossibility to speak about the set of all ordinal numbers, but only about sets of a specific type. The set of these ordinal numbers is, however, a set of a higher type and therefore its ordinal number does not belong to it. Can one, now, speak in 2 about the set of all things of a certain type? Its cardinal number is the greatest for this type, still there exists in the next higher a greater one. Antinomies 3 and 4 are resolved by the remark that the proposition by which a set belongs to itself, or a property applies to itself, respectively, is under all circumstances meaningless. For the type of an object to which a property φ can be meaningfully ascribed is always different from that of φ itself, and no property can therefore be meaningfully stated of itself.

The first hint at a type-theoretical distinction is found already in Peano who made a distinction, in direct contrast to the algebra of logic, between the relation of "being an element" and "being a subset." The justification was that the second one is transitive, the first one not. A further distinction was between an object a and the class ιa that has this object as its only element, with the indication that in case a itself is a class, it can consist of several elements, whereas ιa has just one element. Frege formulated then in a thoroughly clear way the type theory for functions and concepts, with all the rules that belong to it, and this even before any antinomies were known; which should be securest proof for this not to be an *ad hoc* solution to the antinomies. He went even as far as claiming that the relation that obtains between an object and a concept under which this object falls, should be different from the corresponding relation between concepts of first and second order. He made, however, into nothing the effect of this incisive distinction for deterring antinomies by ascribing to each concept an object as a scope, through the function $\varepsilon f x$.

$15\frac{1}{4}$

24-404

Russell cites similarly the justification that Frege gave for his distinction between object (individual) and concept (property). Namely, it states that concepts, in contrast to objects, are incomplete (amenable of addition), whereas Russell says in the *Principia Mathematica* that a propositional function φx is an equivocal symbol because of the variable x that occurs in it, and that this equivocality has to be somehow eliminated in a meaningful proposition (e.g., through the all-sign). Such a condition is not satisfied, however, if one puts a propositional function in the place of an individual in a proposition. (Therefore a propositional function could never occur in the place in which an individual can occur.)

A new presentation of the simple theory of types is found in Carnap's *Logistik*.

The type theory discussed so far excludes all of the logical paradoxes (this is the only thing of concern in the setting up of logical-mathematical systems), and limits on the other hand the set-theoretical ways of inference only so little that the logical foundation of classical mathematics remains possible. Russell does not halt at it, but puts as a basis in the *Principia Mathematica* a refined theory, what is known as ramified type theory, and this happens on two grounds:

1. He wanted to incorporate the antinomies of the second kind in his solution.

2. He believed to have found a (logical) principle that each correct logical system has to satisfy, but that the system of simple type theory doesn't satisfy. This principle, known as the vicious circle principle, is as follows:

> "No totality can contain members that can be defined only with the help of this totality."

If a totality apparently contains such members, it is thereby proved that the totality in question cannot exist. By a totality is here understood a concept for which it is meaningful to speak of all the things that belong to it. It should now be the case that all the paradoxes (cf. the scheme of paradoxes given above on p.) arise from the formation of such illegitimate totalities. The set M of all definable real numbers, for example (cf.), is such an illegitimate totality, because even the diagonal sequence of all finitely definable numbers would belong to it, one that clearly can be built only through a reference to the totality M. The case is analogous for the set of all propositions, in reference to Antinomy $N^{\underline{o}}$ ⟦8⟧.

$16\frac{1}{4}$

24-405

Even the set of all properties of a definite type, as admitted in the simple theory of types, is such an illegitimate totality, because one can define certain concepts of the first order in reference to all concepts of this type. This happens, for example, with the natural numbers, because the property "x is a natural number" is, admittedly, defined through x having *all* hereditary properties that 0 has, and it is itself such a property.

Poincaré in was the first one to declare these so-called impredicative definitions as erroneous. Thereby he rejected the *Frege and Dedekind-Russell* justification of the inference of complete induction. Russell joined this through the rejection of impredicative definitions by the vicious circle principle.

Carnap ( ) shows especially clearly that there is in fact a circularity at hand here. Namely, the question whether the specific number 5 is inductive is made by definition dependent on the question whether 5 has all the hereditary properties that belong to 0. This question can be, however, decided only if one already knows whether the property of inductiveness belongs to 5, which is even such a property. Also Weyl, *Symposion*, p. 13.

It turns out, though, that this kind of definition is very common. It is, especially, indispensable for the fundamental theorem by which each bounded set of real numbers has a lower limit. If one conceives of the real numbers in the Dedekindian way as sections (i.e., as certain classes) of rationals, then the lower limit k of a set M is defined by the condition that a rational number belongs to a subclass of k if it belongs to the subclass of all numbers *from* M. The number thus defined can, on the other hand, itself belong to M. The proof by Cauchy of the fundamental theorem of algebra, for example, would become impossible, something Zermelo (*Annalen* 65) presents as an argument against Poincaré. Poincaré (*Acta* 32) shows in return that one could avoid impredicative definitions in this case by considering the lower limit of a set of rational numbers.

$17\frac{1}{4}$

24-406

One can even show, without the use of impredicative definitions, that each sequence of real numbers has an upper limit, because here the all-sign needs to refer only to natural numbers (the numbering of the sequence) (cf. Weyl *Kontinuum*, p. , Hilbert).

By the vicious circle principle, the functions of a definite type form an illegitimate totality. Therefore, each such type must be split into a series of subtypes (orders), by which propositions about all things of a type of functions become impossible. Russell arrives in this way into a logical system that satisfies the vicious circle principle. He assumes there to exist in the world certain objects $a, b, c \ldots$ (individuals) not further analyzable, and certain concepts not further analyzable (atomic functions), i.e., properties of such individuals, and relations $R_1 R_2 \ldots R_3$ between them (say x is red, x is a relative? of y). The simplest propositions (atomic propositions) have then the form $R_1(a)$ $R_2(bc)$ etc, i.e., they state that a certain property holds and that a certain relation holds, respectively, for some determinate individuals. One obtains from these more complicated propositions through the application of the operations of the propositional calculus, ($\vee \quad \bar{} \quad .$), for example, $\sim R_1(a) \vee R_2(ab)$. If one substitutes an individual by a variable in such, one obtains *elementary propositional functions*, $\sim R_1(x) \vee R_2(ay)$. If one binds all, or a part, of the variables by quantifiers in such, one obtains the propositions of the first order, and functions of the first order, respectively, e.g.,

$(x)(\exists y) \sim R_1(a) \vee R_2(ab)$ and $(x) \ldots$.[12] Propositions (functions) of the first order don't speak about any other totalities than that of the individuals. They form themselves a totality of a higher order.

One can now build further functions (propositions) that speak about the totality of propositions (functions) of the first order,[13] which

$18\frac{1}{4}$

24-407

finds its expression in that there occur variables (φ) the values of which are functions (propositions) of the first order, for example:

$$(\varphi_1) \,.\, \varphi_1(a) \supset \varphi_1(x) \qquad (\varphi_1) \,.\, \varphi_1(a) \supset \varphi_1(b)$$

By the vicious circle principle, these functions (propositions) cannot belong to the functions of first order. They make up for a new type of propositions and functions of the second order, which gives reason to the rise of new variables φ_2 and propositions (functions of the third order). Just as with functions of the first type in the earlier sense, i.e., those that have individuals as arguments, they appear here as divided in all possible orders. (For one can build properties of individuals, in the definitions of which one talks about functions of arbitrarily high order.) For example, $(y)R_2(yx)$ would be one of order 1, $(\varphi_1)\varphi_1(a) \supset . \varphi_1(\hat{x})$ instead one of order 2.

A function must always have a higher order than each of its arguments. Russell calls a function predicative when it has the lowest order still compatible with its arguments. At another place, Russell calls predicative those functions in which there occur no bound variables at all. So for example is predicative, but not.

One could think that one obtains the functions of the first type in the earlier sense, when one combines all of the orders G thus built. This is not at all the case because one can build new properties of individuals that refer to the whole of G and therefore don't belong to it. So, one would not obtain anything more than a continuation of the orders into the transfinite. Russell, by the way, found such a continuation *incorrect*, in opposition to *König*.

[12] ⟦Most likely $(x)(\exists y)(\sim R_1(x) \vee R_2(ay))$ and $(x)(\sim R_1(x) \vee R_2(ay))$ are meant.⟧

[13] Individuals count as of order 0.

Even the functions $\vee\ \bar{}$ of the propositional calculus and the concept "true" (x is a true proposition) decompose in ramified type theory into infinitely many different orders, by the propositions to the order of which they refer.

To show how the antinomies of the second kind are resolved according to this theory, let us consider as an example $N^{\underline{o}}$ ⟦9⟧, cf. ⟦Hilbert-Ackermann, p. 95⟧. Here one has to (for a logistic condition to be possible), make in the first place precise the unclear term definition for a, namely that a property (propositional function) is to be understood under it that applies to the number a, and only to it (a characteristic property).[14] Therefore one cannot speak of all definitions of real numbers, but only of those of a specific order k, and the one built after the diagonal procedure refers to all of the order k and has therefore a higher order.

Ramified type theory has to fight with two difficulties:

1. It makes apparently impossible the formulation of the basic laws of logic, because these treat for the most part with illegitimate totalities (the law of excluded middle, for example, states that each proposition is either true or false). Russell overcomes this difficulty by what is known as systematic ambiguity. It states that the logical basic signs $p \vee q, \sim p$ ⟦added: and most of those introduced by definition⟧ are ambiguous symbols that can designate each of the different types of "or" and "not." The formulas in which these symbols occur denote, correspondingly, not a specific proposition, but they are symbolic forms, i.e., they give an instruction on how one can build infinitely many propositions, by the choice of one among the different meanings of the basic symbols.

⟦2.⟧ The second essential difficulty of the ramified theory of types comes from the impossibility to speak meaningfully about all functions (sets) of a specific type. This has the consequence that one cannot speak of all real numbers, either, but just of the real numbers of a specific order. Therefore the Weierstrassian theorem about the existence of an upper limit holds for no order, because it belongs in

[14] ⟦Many changes have resulted in a complicated phrase. The original contains: One has to determine in the first place what is meant by definition here⟧

general to a higher order as the set the limit of which it would be.

One cannot even define the concept of an inductive number in the above way but has to distinguish between inductive numbers of different orders, by the order they receive in their definition. One could use the inference of complete induction for inductive numbers of the order n only for properties of the order n. Then, for example, the recursive proof that the sum of two inductive numbers of order n is again an inductive number of order n would already be impossible.

These complications make impossible a large part of the proofs of classical mathematics. To avoid them, Russell puts up his dubious

** Similarly of course for all types that follow.[15]

$19\frac{3}{4}$

24-408

axiom of reducibility which states that there exists for each function a predicative function with the same scope. In mathematics, the question is always only about the scope (the extensional character of mathematics), and therefore the axiom has as a consequence that one can limit all considerations to predicative functions. These build, now, a legitimate totality, and the ramified type theory is thereby in practice reduced to the simple one.

Russell (   ) could present as a justification for his assumption of the reducibility axiom only that it leads to the desired consequences in mathematics, and to no others. It is a justification in the sense of (p.   ). (For more about this, cf. . .   .)

This idea of Russell's is reflected in the following logical system, specified in the *Principia:*

Undefined ground symbols:

1. Variables for propositions $p\ q\ r \ldots$
2. Variables for individuals $x\ y\ z$
3. Variables for predicative functions of order 1, notation $\varphi!\ \psi!\ \chi!$

   order 2, notation $f'\ g'$

---

[15] ⟦Not clear where this footnote belongs.⟧

and so on for higher levels. To indicate the kind and number of their arguments, the latter are equipped with a ^ and set at the end, e.g.,

$\varphi!\hat{x}, \psi!\hat{x}\hat{y}, f!(\hat{\varphi}!\hat{x}, \hat{y})$

4. Variables for arbitrary (also non-predicative) functions of order 1, φ, ψ, χ

and order 2, f g etc.

By the above, an all-sign can never refer to such variables, but they are allowed to occur only as free variables (cf.   ) and their meaning is that a correct proposition arises if a function of an arbitrary order is written in their place. It is analogous with propositional variables that are ambiguous in type as well. Variables for non-predicative functions of a specific order are, therefore, because of the reducibility axiom, superfluous.

$20\frac{3}{4}$

24-409

5. $\vee$ or $\sim$ not (typical ambiguity)
6. (x) for all x similarly (φ) (f) (in which φ f is predicative)
7. $\vdash$ assertion sign
8. $(\exists x)$ is introduced as an abbreviation for $\sim(x)\sim$, and for

$\sim p \vee q \quad p \supset q$

Identity is defined in the Leibnizian way, as agreement on all properties, i.e., $x = y$ is an abbreviation for $(\psi)\ \psi!x \supset \psi!y$.

Each meaningful expression of the system (a proposition or a propositional function) arises by starting from the propositional variables $p, q, \ldots$ and elementary expressions of the form

$\varphi x!x,\ \varphi x,\ \varphi xy,\ f!(\psi!\hat{x}, y)$ etc

(i.e., function variables the argument places of which are filled by function variables of the corresponding types), and by applying the operations 5, 6 arbitrarily many times, though just on predicative variables.

The axioms are ⟦*Principia*, part I, section $A{*}1$:

$*1\cdot 1$. Anything implied by a true proposition is true.

$*1\cdot 11$. When ϕx can be asserted, where x is a real variable, and $\phi x \supset \psi x$ can be asserted, where x is a real variable, then ψx can be asserted, where x is a real variable.

$*1\cdot 2$. $\vdash : p \vee p \,.\, \supset . \, p$

$*1\cdot 3$. $\vdash : q \,.\, \supset . \, p \vee q$

$*1\cdot 4$. $\vdash : p \vee q \,.\, \supset . \, q \vee p$

$*1\cdot 5$. $\vdash : p \vee (q \vee r) \,.\, \supset . \, (p \vee q) \vee r$

$*1\cdot 6$. $\vdash :. \, q \supset r \,.\, \supset : p \vee q \,.\, \supset . \, p \vee r$

$*1\cdot 7$. If p is an elementary proposition, $\sim p$ is an elementary proposition.⟧

In each of the axioms, variables have a typical ambiguity. For example, ⟦$*1\cdot 11$⟧ contains the case that x itself is a function of first order $\varphi!\hat{x}$, by which one gets

$$(\varphi) f(\varphi!\hat{x}) \supset f(\psi!\hat{x})$$

There would be as rules of inference:

1. Analogously as in Frege.[16]

2. Free variables for individuals and predicative functions can be turned into bound ones through an all-operation set in the beginning of the formula.

3. The rule of substitution, incidentally not explicitly formulated by Russell, reads as: Arbitrary meaningful expressions can be substituted for propositional variables, arbitrary functions of one argument of the appropriate type and built from the basic signs of the system can be substituted for function variables (with the condition that no overlap of scopes occurs with variables equally denoted).

Russell proposes two possible axiom systems, with (x) and $(\exists x)$ as basic concepts, but with $\vee$ and $\sim$ used in a first instance as basic concepts only for the type that refers to propositions without bound variables (elementary propositions), the other types of $\vee$ and $\sim$ introduced through the following definitions:

---

16 ⟦A cancelled passage has: 1. The rule of implication.⟧

$$p \vee (x) q\, x = (x) \,.\, p \vee q x \qquad \text{Df}$$

$21\frac{3}{4}$

24-410

$$\sim (x) \varphi x = (\exists x) \sim \varphi x \qquad \text{Df}$$

Analogously for $(\exists x)$ in place of (x).

The application of these rules allows to place all the quantifiers of an expression in front, so that their scope becomes the whole expression, and $\vee, \sim$ etc refer now to elementary propositions.

Axioms 1–5 are now essentially weaker, namely assumed only for elementary propositions (which amounts to allowing only formulas without bound variables to be substituted in them), and therefore certain substitution axioms become necessary. This can be found carried through in (   ) and in Herbrand.

As regards definitions, Russell is more liberal than Frege to the extent that he doesn't necessarily require an explicit definition for newly introduced signs, in the form of an equation, but just a statement of how one can translate back propositions in which the new sign occurs into ones in which it doesn't occur anymore.

Symbols introduced in this way are called incomplete and this kind of definition is used (analogous with Carnap, Dubislaw). Russell succeeds in this way to define two of the previous basic signs, $\varepsilon\varphi x$ and $\backslash x$. The sign $\backslash$ appears in Russell in the form $(\iota x)\varphi x$, the x that has the property φ. The sense of the proposition about *the x for which* is established by the following definition:

$$\Psi[(\iota x)\varphi\, x] = - -$$

Then, if the x with the property φ doesn't exist, i.e., if there exist either no or several x with the property φ, each proposition about it becomes false.

The concept of scope, $\hat{x}\varphi x$, is introduced analogously through the foregoing usage, for

$22\frac{3}{4}$

24-411

guaranteeing that concepts under which the same objects fall have the same scope, which is the essential property.

Classes appear, then, only as *façon de parler* (logical fictions) in the system, and they are always eliminable (no class theory).

In the derivation of arithmetic from the above axioms, one has to observe first that there exist infinitely many different types of inductive numbers. For example, cardinal numbers that correspond to sets are different from cardinal numbers that correspond to sets of sets. This does not disturb, however, because the same propositions hold for all of these. We have namely in general that the validity of a proposition is maintained when one raises by the same number all of the types that occur (this holds for the axioms). The same does not hold in general with type lowering, for one can show, for example, that there must exist at least two classes[17] but not that there exist at least two individuals.

It is often important in arithmetic that the propositions be maintained when the types are raised in different ways in different parts. This, though, does not hold in general (counterexample **?** p. 53, *Principia II*, p. **?**), and quite complicated conventions are therefore needed (cf. *Principia*,) for interpreting the non-ambiguity of symbols, i.e., for taking, as a substitute, for certain sets of propositions a single form, not anymore ?? ambiguous.

For the rest, the theory of finite numbers lets itself be developed just as in Frege, up to the trouble that type theory makes impossible the proof of existence of infinite sets and therefore cannot give, either, a proof for the proposition that there exists for each finite number a greater one in the same type, because the Bolzano-Dedekind proof produces a relation between objects and concepts. These are, however, two things of different types and therefore they don't yield any mapping to themselves.

In the Fregean proof, the cardinal number of the set of natural numbers from 0 to n has a type different from these numbers. One can

---

[17] ⟦A tiny incoherent note is added above: (types exist (that consist only of (that consist of one element)⟧

still prove the existence of arbitrarily great numbers, by stepping up to ever higher types. For it follows from the axioms that there exist at least 2^n classes of the n-th order. This, though, is not sufficient for all purposes, because the quantifiers are always allowed to refer only to a specific type. One needs therefore, for the theory of real numbers, a new axiom, what is known as the axiom of infinity that Russell formulates as follows: There exists for each inductive number a class of individuals to which this number applies, i.e., the same as: The set of individuals is not inductive.

$23\frac{3}{4}$

24-412[18]

(Which requires less than that it be reflexive.)

Axiom systems of set theory don't have any type distinctions, and one can carry through the mentioned proof of the existence of infinitely many things and to give a foundation for the theory of natural numbers without the axiom of infinity, cf. von Neumann, *Math. Zs.* 27. One needs a proper axiom of infinity also in set theory only for the proof that the infinitely many natural numbers form a set.

Another difficulty for the building up of mathematics from the Russellian axioms is brought by the axiom of choice that likewise cannot be derived but that on the other hand is required for the proofs of many theorems. This is especially the case for calculation with transfinite numbers. The theorem by which the sums of equivalent sets are equivalent, needed already for the definition of the sum of cardinal numbers, depends in the case of infinitely many summands on the axiom of choice. Tarski gives in (*Fundamenta*, vol. V) a series of theorems from the theory of cardinal numbers that are equal to the axiom of choice, including $m = m^2$, $mn = m + n$ (for transfinite m, n) and the monotonicity of sum and product for transfinite numbers.

The axiom of choice is needed already in the theory of finite sets. The identification without the axiom of choice of the two definitions of finite sets, by Frege and by Dedekind (inductive and non-

[18] ⟦The page count number at bottom is off by one.⟧

reflexive), follows only halfway, namely that each inductive set is non-reflexive. A great number of other definitions of the concept finite were put up ever since, surveyed by Tarski in ( ). For example, a set is called finite: **1** If it can be doubly well-ordered (Stäkel [1907]) **2** If it has under each ordering a first element. 3. If there exists a Zermelo mapping, i.e., a unique mapping into itself that maps no proper parts onto it ⟦a heavily cancelled passage follows⟧ including its definition that appears to be, without the axiom of choice, equivalent to neither the one of Dedekind nor the one of Frege. Tarski ( ) gives five[19] that can be ordered according to their strength, presumably identifiable without the axiom of choice.

The following two stand between Frege and Dedekind: $\mathfrak{m}$ is called finite if

1. Each growing set of subsets has a most comprehensive element.
2. The power set is irreflexive.

(If one requires irreflexivity of the power set of the power set, that is as Russell shows equivalent to Frege's definition.) A definition that requires apparently even less than that of Dedekind reads: A is finite

$23\frac{3}{4}$

24-413

if it is not the sum of two disjoint sets of the same cardinality as A.

Independence of the axiom of choice, the work of Sierpinski.

Derivation of mathematics, set theory not included.

One should not describe the execution of the logistic program by Russell as an entirely satisfactory one, because certain undoubtedly non-tautological axioms are required for the foundation of mathematics (infinity and reducibility, for several theorems also the axiom of choice). This shortcoming comes especially clearly out when one grasps more clearly the essence of what is logical and what the concept of a tautology. It became thus completely clear that the re-

[19] ⟦A heavily cancelled passage suggests a series of five definitions.⟧

ducibility axiom is an extra-logical condition, because one can construct worlds (the system of individuals and atomic functions, cf. p.) in which it does not hold (cf. Wittgenstein , Ramsey , Bernays). Whether it holds in our world is an empirical question.

As concerns the axioms of choice and infinity, Russell had already acknowledged the above in ( ), when he introduced these as explicit conditions in theorems in the proofs of which they were used. He went later even further by not counting as a tautology the theorem, provable in the *Principia*, by which there exists at least one individual and therefore two in classes of order n. One can get through[20] these difficulties by making it clear that it is not the theorems of mathematics themselves that are of a logical (i.e., tautological) nature (as Frege had believed), but just those implications that state that they follow from some extra-logical conditions (return to Peano).

This point of view was represented by H. Hahn in ( ). He referred to the fact that the concepts of the Russellian system, individuals

$24\frac{1}{4}$

24-414

and atomic functions, have no absolute meaning but can be realized in different ways in the world, and that the theorems of mathematics hold only for such realizations in which certain "substantial conditions" are satisfied, from which they follow through tautological conversion, i.e., logical inference.

The axioms of choice, infinity, and reducibility are existential axioms. Therefore they require in a certain sense a richness of the domain that is laid as a foundation. Such realizations exist neither in the *Brouwerian* constructive nor in the empirical sense, but one arrives at them through idealizations of reality, similarly to, say, Euclidean geometry. The freedom from contradiction of the requirements of richness remains an open question.

Ramsey represents a contrary point of view (tied to Wittgenstein's

---

[20] ⟦This word is cancelled.⟧

line of thought). He explains that all of the theorems of the *Principia* (as well as the axioms of reducibility and choice) become tautologies if one just gives to the ground symbols another intepretation, changing them in the sense discussed on p. . Russell had understood by "propositional functions" those finite expressions one can construct by certain processes, starting from the atomic functions. One is necessarily led to the distinction of different orders. Russell takes this to be an interpretation, because the concept of a function becomes thereby not objective, but to a certain extent "anthropological," determined by the means of expression that are by chance humanly available. In particular, the difference between functions of different orders lies merely in the way in which they are represented linguistically, not on them as such.

The concept of function (and the concept of set) becomes, further, too narrowly determined here. There certainly exist "lawless" sets that don't allow at all of a construction (i.e., of a representation in a finite form), because there exist undenumerably many sets but only *denumerable* many constructible ones (cf. on this).

$25\frac{1}{4}$

24-379R

What Ramsey sets in the place of the Russellian development is the following: He declares predicative functions to be expressions that arise from the atomic functions through finite *or infinite* iterated application of the operations of the propositional calculus. Here the circumstance that one cannot write down infinite expressions is set aside as a casual fact that concerns the empirical and is of no concern for logic. The domain of functions thus declared has the property that one obtains nothing more by the consideration of new functions. Thereby the axiom of reducibility becomes a tautology and ramified type theory superfluous (simple type theory is maintained). The impredicative functions lose their circularity, because they are to be considered, not as production principles for functions but merely as characterizations of certain functions already at hand within the domain of predicative functions, in consideration of this domain, something that is as little circular as the definition of a person through the property of being the tallest in a certain group.

One step further in this direction leads Ramsey to the definition of functions *in extenso*, by which he understood an arbitrary association of propositions to individuals, irrespective of whether this association can be singled out through a finitary law or not. If one accepts this concept, then even the axiom of choice becomes, says Ramsey, an obvious tautology, because no law is required now for the choice set that would choose it from the individual elements.

Ramsey's attempt at an improvement of the *Principia* depends in the end on a dogmatic treatment of infinite sets, with which one operates precisely as with the finite ones. It is clear that the difficulty of choice and of the reducibility axiom disappear then, because these are satisfied in a trivial way for finite domains of individuals. Ramsey has, though, still the merit of having carried this point of view, (one could call it infinitistic, Carnap speaks of a theological one), through in a consequent way.

24-415

As concerns the axiom of infinity, Russell claims that it must be either a tautology or a contradiction, not an empirical proposition. He justifies this through a conception about the essence of identity that deviates from the Russellian one and goes back to Wittgenstein. (The Russellian definition is modified because it is logically possible that two different things agree by chance in all properties, cf. Langford.)[21]

Russell's views on antinomies of the second kind are more noteworthy. In the absence of the ramified theory of types, he has to give a new resolution for them.

26

24-416

He was the first one to establish that there occurs in all of these the relation between a symbol and its meaning (the relation "*a* means the object *b*"), and he sees the cause of the antinomies in the impossibility to build (define) such a relation in its generality. One

[21] ⟦Langford: Otherness and dissimilarity, *Mind*, vol. 39.⟧

must instead distinguish different orders in it, by which a symbol in which there occurs a relation of order n to an object that it designates, stands in a meaning relation of order $n+1$. The antinomy of the heterological disappears, for example, because one has to define accordingly: "A property word is called heterological of order n if it stands in a meaning relation of order n to one of its properties." The words heterological of order n don't stand, instead, in a meaning relation of order n to any of their properties (but just to a certain property in a meaning relation of order $n+1$).

The ramified theory of types turns back by the above, to the extent that one can divide expressions for functions and propositions in orders after the order in which they denote their objects. These divisions refer, however, only to the linguistic representation of the functions and propositions, not to themselves (for objects can be denoted by symbols of different orders).

Carnap( ) develops other ideas (not presented in detail) to show the impredicative definitions as allowed and the ramified type theory as superfluous. He recognizes that the all-sign should not be extensional in logic (it does not mean "for each single"), but it means "logically derivable" by which an all-sign that refers to functions does not presuppose the totality of functions (there is a similar idea in Langford).

Russell maintained in the second edition of the *Principia* the ramified theory of types, but put the reducibility axiom aside, as a non-logical proposition. He put in its place another one (held to be purely logical), namely the extensionality thesis. A propositional function the arguments of which are propositions (e.g., $p \vee q$), is called extensional if its truth value depends only on the truth values of its arguments. A propositional function with propositional functions as arguments is analogously extensional if its truth value depends only on the scope of the argument functions (as in, e.g., $\varphi(x) \supset_x \psi(x)$).

All functions that occur in mathematics are extensional, and extensionality states that each function is extensional, or, in a sharper form (Russell), that each function can be obtained from the atomic functions through the application of

27

24-417

the operations ($\overline{\ }$ $\vee$ ( ) $\exists$).

Extensionality follows by itself from the conception of logic that was described on page (Wittgenstein), and was also laid as a basis by Ramsey. Numerous examples seem to contradict it. For example, A believes p, q follows from p, proposition p speaks about the object a, and analogous propositions for functions, for example, A believes that $(x)\varphi x$ is true, are clearly not extensional.

Proponents of the extensionality thesis maintain that these examples speak only apparently about the proposition p, in reality instead rather about the conceptions of a person about certain objects that occur in p, or even about the symbols (characters, words) through which p gets expressed. Russell goes in this direction, a discussion of this question, not entirely clear, in appendix in which he explains that the p in A believes p and in $\sim p$ denote two completely different things, factual property and asserted property. Frege had already claimed this, by his distinction into the sense and meaning of a proposition ( ). Different forms of assertions equivalent to the extensionality thesis (for propositions) are found in ( ), for example ⟦missing⟧

Here V and F can denote arbitrary true and false propositions.

It becomes possible to identify classes and functions by the extensionality thesis for functions, and the definition of $\hat{x}\varphi'$ on p. becomes superfluous.

Russell expects a partial substitution of the reducibility axiom (if one expresses it in the above sharp form - *even possible to consider it a definition*), something that is confirmed in the concept of identity. Namely, it follows by its help that if two functions coincide on all predicative properties, they coincide also with the rest. (The reducibility axiom was used earlier for that.) Russell claims further that it is sufficient also for the theory of natural numbers and gave a foundation for that in appendix B, in which he tried to show that the inductive numbers

$1\frac{1}{2} + 28$

24-418

of order five are identical to those of each higher order.

The extensionality thesis suffices in no case for the theory of real numbers, as Russell himself emphasizes. One sees even from Weyl's *Kontinuum* how far one can build up analysis without the reducibility axiom. There, however, the natural numbers and the pertinent ways of inference (complete induction) are assumed as given.

Chwistek attempts in   a construction of the elementary concepts, under the assumption of ramified type theory but without the axioms of reducibility and substitution. He arrives as far as proving the Schröder-Bernstein theorem and the closure of inductive numbers of a specific order with respect to addition, multiplication, and exponentiation. Chwistek suggested, for the further construction of mathematics, to add the axioms of semantics, i.e., he incorporated the concepts and propositions that one otherwise treats in metamathematics in word-language, expressly in his theory. He describes the formal system that arose in this way in   . It turns out that there would exist for each type denumerably many functions (proof of the axiom of infinity). Further, the axiom of choice holds, because only subsets ⟦no continuation⟧

The question that concerns the construction of mathematics without the reducibilily axiom belongs to ones very little clarified in foundational research. Moreover, the few results at hand are mostly presented in a form in which one cannot consider them definite.

24-419

Axiomatization imprecise because of logic, this is overcome through formalization.

24-453

1. Burali	Φ ordinal number, *f* the next greater
2. Greatest cardinal number	Φ cardinal number
3. Russell	set of alls sets
4. Impredicativity	of all concepts
5. Relation of all relations	of all relations

---

6. Liar
7. Heterological
8. Set of all propositions
9. Smallest number with 100 words
10. Smallest ordinal number
11. Richard

24-454

⟦4.⟧ Clear version, from the beginning to antinomies

24-455

Logicism

One understands by logicism the conception, going back to Leibniz and established by Peano, Frege, and Russell, that mathematics is reducible to logic, i.e., that mathematical concepts are definable from logical ones and mathematical theorems provable from the axioms of logic.

As concerns the precise sense that one has to give to the term "logic" with this claim, one is content, to start with, with a listing of certain concepts and basic propositions that one tends to describe as logical in the common way of speaking. It is only in the very recent times, especially under the influence of *Wittgenstein*, that supporters of logicism have undertaken attempts at clarifying the concept of logic, i.e., to indicate the features by which logical concepts and propositions differ from others (Russell   , Hahn   ). In these, the characteristic of the logical concepts (also called logical constants) is that they are not signs for anything that subsists in the world, but just serve to express the form of sentences. They reveal themselves only in the way in which language associates symbols to reality.

24-457

The possibility of logical propositions is brought about by this association not being one-to-one, so that the same state of affairs can be expressed in different ways (for example, through p and $\sim\sim p$). A proposition that ⟦expresses⟧ the fact that p and $\sim\sim p$ are associated to the same state of affairs (i.e., $p \equiv \sim\sim p$), is then a theorem of logic, and similarly a proposition which states that a certain combination of symbols (e.g., $p \vee q$) is associated to a more comprehensive class of states of affairs than another, p. (I.e., in our case, from p follows p or q). It follows from this conception that the logical propositions don't state anything about reality, but result in truth already from the way in which the symbols are associated to reality.

To prove the thesis of logicism it would be absolutely necessary, as Frege already remarked, to replace the imprecise and ambiguous word-language through an exact formula language. It is only through such an exact language that one can specify the logical basic propositions from which one starts, and the logical rules of inference by which one derives further propositions from the basic propositions, to such an accuracy that the unnoticed creeping in of extra-logical elements during deduction is avoided.

There results as a characteristic of the rules of inference that they are purely formal,

24-459

i.e., relate only to the external shape of formulas, so that they can be applied mechanically (without knowledge of their meaning). The carrying through of the logicist point of view gave, then, (independently of its reaching or not reaching its aim) the result by which one can express mathematical propositions through combinations of a few basic symbols, and deduce them from a few basic propositions by a few mechanical rules.

The founders of the algebra of logic (Boole, Peirce, Schröder) offered a groundwork for this so-called formalization of mathematics. It was, however, only Peano who gave to the logical symbolism the flexibility that is needed for the expression of even complicated mathematical propositions to be possible. He and his coworkers presented in the [*Formulaire*] number theory, arithmetic, analysis, geometry in a purely symbolic form, one in which the propositions and concepts of the field at hand were led back to a minimum of unproved propositions and undefined basic concepts. The reduction of these latter elements to logic was, instead, not aimed at.

Peano founded the arithmetic of the natural numbers on three concepts, "number," "0," and "successor" (where by the successor of x is to be understood the number $x + 1$). He stipulated five axioms for these three concepts, ones that state that each number has exactly one number as successor, that 0 is not the successor of any number, that each number distinct from 0 is the successor of exactly one number,

24-461

and finally that a property which belongs to 0 and that belongs to the successor of x whenever it belongs to x, belongs to all numbers (axiom of complete induction).

One finds an improved formulation of the axioms of Peano in [ ], and a logical derivation of arithmetic, not yet entirely unobjectionable in Peano [cf. definition by complete induction], in [ ].

As concerns rigour in deduction, the *Formulaire* does not satisfy today's requirements anymore.[22] The rules of inference and of definition, especially, are not formulated sufficiently precisely.

The execution of the logicist program was for Peano by no means the main aim. He wanted rather to create, in the first place, an encyclopaedia of proved mathematical theorems (together with historical remarks), something that is possible in a relatively small space, because of the shortening effect of the formula language of about 1 to 10. Peano hoped also to make, through the introduction of an appropriate symbolism, a step ahead in the ease of mathematical inference, such as was achieved in calculation with numbers through the decimal system.

Frege delivered the first execution of the logicist program in [⟦1879⟧]. He achieved a precision in the formulation of his basic propositions and rules of inference that still satisfies to the full the most modern requirements,

24-463

one that was reached again only by the formalistic school and Polish metamathematics. The logical basic concepts that he laid as the undefined basis of his deductions are the following ones:

1·) "Object." This concept has no special sign but appears only through variables a, b, c that denote arbitrary objects.

2·) "Function." By this is understood an arbitrary unique association of objects to objects, i.e., a procedure f through which an object b is associated in a unique way to each object a (b is called the

---

[22] ⟦The first draft has: Cf. Frege, letter to Peano, *Rivista di Matematica*, VI, p. 53.⟧

value of the function f for the argument a). The symbolic expression for a function consists of a combination of signs that contains an object variable a and that designates a determinate object when a is replaced by a name of a specific object (e.g., the father of a, $sin\, a$). If, in particular, the value of a function f is a proposition (as, for example, in the function expressed by a is red), it is called a "propositional function." Functions that are not propositional functions are called "object" functions, or descriptive functions.

24-465

Frege allowed also of functions with several arguments (e.g., $a+b$, $x>y$). Propositional functions with one (and several) variables are, as is apparent, expressions for properties and relations. Frege counted functions not as objects, with the justification that they are incomplete (amenable of addition), namely through the substitution of a name in an argument place. Even the concept of a function has no special place in the system, but occurs only through the sort of variables $f, g, h \ldots$ that denote arbitrary functions.

3.) F-Function. Designated through a special sort of variables F, G, these are functions the arguments of which are not objects but instead functions.

The remaining Fregean basic concepts are the following:

1. Negation
2. Implication. [This is the concept "from a follows b"]
3. Identical with
4. All
5. The object such that
6. "Course-of-values of a function." This is what is common to two functions that always assume the same value for the same arguments. For example, $sin\, 2a$ and $2\, cos\, a\, sin\, a$.

Frege considered the value of a propositional function $\varphi(x)$ for the argument a to be, not the proposition $\varphi(a)$ itself, but the truth value of this proposition,

24-467

and then the course-of-values of a propositional function is what one usually ⟦calls⟧ the "scope of a concept" or "set of objects that fall under a concept." The course-of-values of a function was counted, contrary to the function itself, as an object, which had ill-fated consequences in regard of the antinomies.

Each of the above three concepts has a special sign in the Fregean system, and how the meaningful expressions are built up from these signs, with the three sorts of variables, is defined with painstaking care (grammar of the formula language).

Eight formulas are put on top as logical axioms and . . rules of inference formulated, by which all deductions are derived in Frege's work.

Frege sees the essence of a definition in the introduction of a simple name in place of an expression composed out of the above basic symbols. He was also the first one to formulate in a precise way the rules by which this has to take place. It was also his merit to have been the first one to distinguish between those assertions (concepts) that can be expressed exclusively in the logistic symbolism and those that relate to the symbols of the system and have to be formulated in ordinary word-language (Hilbert's distinction between mathematics and metamathematics). To the second category belong, for example, the definition of a

24-469

"meaningful expression," as well as the formalization of the rules of inference.

Frege used his system in the first place to develop a purely logical theory of natural numbers (later rediscovered by Russell). It relies on the concept of equivalence between sets. (Two sets M, N are called equivalent whenever their elements can be associated to each other in a one-to-one way, i.e., whenever there exists a relation R of the kind that there exists to each element of M exactly one element of N that stands to it in the relation R, and to each element of N exactly one element of M to which it stands in the relation R.) Cantor had already defined the cardinal number of a set as the general concept it has in common with all equivalent sets. This is what is known as a definition by abstraction, the essence of which was likewise analyzed first by Frege: That to each thing x of a domain B (in our case the domain of sets)

24-471

in which a reflexive, symmetric, and transitive relation[23] is defined (in our case the relation of equivalence), a new thing $\varphi(x)$ is associated (in our case the cardinal number of x), in the way that $\varphi(x) = \varphi(y)$ if and only if x stands to y in the relation R. Other examples of definition by abstraction are the concepts of direction (R = parallel), of shape (R = similarity), and of a vector (R = parallel in the same direction and the same length). This definition is, apparently, no mere giving of names but creates new things (the $\varphi(x)$). Still, according to Frege and Russell, one can subordinate it to the scheme of a usual nominal definition, if one understands by $\varphi(x)$ the set of all things from B that stand in the relation R to x. One arrives in this way to the definition: A number is a set that consists of all the sets that are equivalent to a set. But the essentially new achievement of Frege and Russell, in comparison to Cantor, consisted in the definition of finite numbers and the logical derivation of complete induction it made possible.

A property is said to be hereditary if, whenever it

24-473

belongs to a number x, it always follows that it belongs to the number $x + 1$. A number is said to be finite if it is an element of each hereditary set that contains 0. The theorem of complete induction follows immediately from this, and the rest of the Peano axioms follow equally easily (in the system of Frege). Frege proved especially that there exists to each number a successor, by showing that the set of numbers from 0 to n has the cardinal number $n + 1$.

Objections to the Frege-Russell definition of natural numbers were

---

[23] A relation is called reflexive when $R(x,x)$ holds, symmetric when from $R(x,y)$ follows $R(y,x)$, and transitive when from $R(x,y)$ and $R(y,z)$ follows $R(x,z)$.

raised above all by Poincaré. He accused it of circularity, as the natural numbers are already presupposed in the basic concepts of logic. One talks about relations with two terms, products of two propositions, etc, to which Couturat justly objected that this apparent circularity should arise from the imperfection of the word-language. It disappears with a symbolic representation, and further, that one should not need the concept of number to think of two objects being in a relation (*Rev. met.* 1906 p 221).

Poincaré's objections to complete induction (cf.) have been recognised. He took it to be a synthetic judgment a priori and saw in it the principle by which mathematics gets distinguished from logic (cf.).

24-475

Dedekind tried in ⟦1888⟧, almost at the same time as Frege, to build up the theory of natural numbers purely logically.[24]

[24] ⟦The first draft continues from this point to the end of the section on logicism.⟧

⟦5. The epistemological standpoint of the logicists⟧

24-477

The epistemological standpoint of the logicists is the empiricist one, nay, one can say it is precisely this standpoint that gives a strong impulse to the logicist foundation of mathematics. For a consequent empiricism has to confront somehow the fact that there exists a mathematics (that appears to contradict it). From the time when the view of mathematics that originated with Mill was seen as inadequate, namely that mathematical knowledge should have resulted through inductive inferences from experience, there remained only the possibility of showing it to be analytic. Thereby the problem about the relation of mathematics to the other natural sciences would become reduced to that about logic and the natural sciences. This latter relation consists, though, by the empiricist conception, simply in that logic deals with the prescriptions of how linguistic symbols are associated to reality. Why such makes necessary a great, comprehensive science, depends on this association being, by no means one-to-one, but very intricate. For one can express the same empirical state of affairs linguistically in many different ways.

The task of logic consists, then, precisely in the determination of when two linguistic expressions convey the same empirical state of affairs, under the conventions by which the symbols are used, or when a linguistic expression A is contained in another B (under these same conventions), i.e., if the empirical state of affairs expressed by A is even stated by B. It is apparent from this that a logical (mathematical) proposition states nothing about reality, but depends merely on the conventional prescriptions by which we associate symbols to reality. Therefore the proposition can even be asserted independently of all experience as, say, $\sim\sim p \equiv p$ with point sets that

24-479

expresses nothing else but that we agree to use the symbol $\sim$ in such a way that the same state of affairs is associated to $\sim\sim p$ as to p.[25]

---

[25] ⟦This paragraph was originally followed by a text that ends abruptly:

The reason why this ambiguous association of language to reality is carried out is the lack of knowledge that makes it necessary to formulate propositions that leave still a great deal of room for reality, something from which the containment and equivalence of propositions results by itself.

It results by itself that nothing new can ever be achieved through mathematical inferences. If it appears otherwise so far (Hertz waves, etc), that is to be explained by the laws of nature, say the law of gravitation, already claiming much more than has been observed so far, and therefore much more is contained in them than was observed so far.

This standpoint was expressed in particular by Wittgenstein in the formulation "that each proposition is a truth function of the atomic propositions." By atomic propositions are here meant those that cannot be analyzed further, the truth (falsity) of which can be determined by immediate observation. Let it be assumed that there are n atomic propositions. (If there were infinitely many, a corresponding consideration would come to play.) If one knew of each atomic proposition whether it is true or false

System of constitution, rejection of metaphysics, execution of this program⟧

⟦6. Logical calculus⟧

24-420

One can consider the formalization of a discipline as a step beyond axiomatization, i.e., one abstracts also from the meaning of the logical signs (therefore from all signs altogether). This has as a consequence that inference can succeed only through specifically formulated rules that don't refer to any meaning of the formulas anymore. The question whether a proposition is provable from some given axioms obtains only hereby a precise and above all objective sense. Some examples of formal systems were already treated in chapters 2 and 3 (Frege, Russell, and systems of the Hilbert school).

As soon as logic itself is formalized, one can easily make a formal system out of each contentful theory. Herbrand shows in detail how this is to happen. Tarski began in the *Monatshefte* with a general theory of formal systems. It starts from the basic concepts "set S of meaningful propositions" (of the discipline in question) and "set of consequences $F(X)$ of a set X of propositions." One understands with S in applications the set of combinations of signs put together in a specific way, to be made precise, from the basic signs of the discipline. By $F(X)$, one understands all those propositions that one can obtain, starting from the set X, by finitely iterated application of the rules of inference.

Tarski presupposes five axioms about S and $F(X)$, ones that are satisfied in all known formal systems, in particular:

$$S \text{ is a denumerable set, } F(F(X)) = F(X), \text{and } F(X) = \Sigma_Y F(Y)$$

In the last formula, Y has to run through the finite subsets of X. The following axiom is, further, needed for some purposes: There exists a proposition of which each meaningful proposition is a consequence (such is, for example, $p\,.\sim p$ in the *Principia Mathematica*).

Tarski draws from his axioms a series of consequences that concern the concepts of "system" ⟦added above: "the basis for a system"⟧, and especially those of "completeness," "axiomatizability," "freedom from contradiction," "independence," further the "degree of completeness."

1

24-421

A set of propositions ⟦added above: M⟧ for which $X = F(X)$ is called a "system." A set of propositions X is called "axiomatizable" if it has a finite "basis," i.e., a finite set of which it is the set of consequences. M is said to be free from contradiction if not every meaningful proposition is a consequence of M. (This concept of freedom from contradiction is independent of negation, and it agrees with the otherwise usual one for all systems in which negation with its usual properties occurs, cf. already: Post .) M is called independent if no proposition of M belongs to the set of consequences of the others. The "cardinal degree of completeness" $g(X)$ of a set of propositions X is the number of the different systems that contain X. X is called complete if $g(X) \leq 2$ ⟦corrected from: 1⟧. (If the propositional calculus occurs within the system, even this agrees with the usual concept which requires that for each proposition, either x or $\sim x$ belongs to the set of consequences.) The ordinal degree of completeness $\gamma(A)$ of M is the smallest ordinal number for which there does not occur anymore a monotonely growing series of systems free from contradiction that contain M. It is shown that $\gamma(A) \geqq \omega$ (or $\leqq \omega$) when - - - - . Further, for the cardinal number holds $\gamma(A) \leqq g(A)$ whenever $\gamma(A) \neq \Omega$.

Different criteria are given for the axiomatizability of a system. It is shown further that each system free from contradiction can be extended into a system that is free from contradiction and complete. To obtain further results, Tarski adds two operations $n(x), c(xy)$ that produce from each proposition x (each two propositions x, y) a new one (negation and implication, respectively). As many axioms are postulated about the concepts $c(xy), n(x)$ as to show that each formula valid in the usual propositional calculus is in the set of consequences of each set (also the empty one). Some results of the theory thus extended are given in ( ), among these that each system has an independent basis and that no system can be represented as a sum of finitely many ones different from it. (This is trivial for axiomatizable systems.)

The range of formal systems that occur in practice is essentially narrower than is delimited by the above axioms. Above all, meaningful propositions occur always as combinations of signs, ones that are built up from signs of four kinds, as presented first by von Neumann in , namely:

⟦1.⟧ Constants (e.g. 0, 1, $\aleph_0$, etc)

⟦2.⟧ Variables $x, y, \ldots$

⟦3.⟧ Operations $O_1(\ ), O_2(\ )$ that yield meaningful expressions if

2

24-422

one substitutes a definite number of meaningful expressions in the place-holders (one writes usually the operations between the expressions on which it is applied, for example +, =, $<$).

4. Abstractions, i.e., variable binding operations, e.g., $(x)(\ )$, $\hat{x}(\varphi x)$ (= the set of x for which φx holds). The historically oldest one is the integral sign $\int_a^b f(x)dx$. Abstraction has the property that it "binds" the variable, i.e., no specific object can be reasonably substituted for a variable to which a sign of abstraction refers.

Each meaningful expression is obtained through a finitely iterated application of the operations and abstractions of the theory in question, starting from variables or constants. To indicate the order in which the operations or abstractions are to be applied, one makes use of parentheses ⟦by which ⟧ $(a \supset b) \supset c$ becomes distinguished from $a \supset (b \supset c)$, or of points (Peano, Russell), $a \supset .\, b \supset c$. Lukasiewicz shows that both are superfluous if the signs for operations are always placed separately in front of the variables or constants on which they are applied. For example, the two formulas above read then as $\supset\supset abc$ and $\supset a \supset bc$. It is easy to characterize how the operations and variables should follow each other for a meaningful formula to emerge.

The propositional calculus is the historically oldest and simplest formal system. It surfaces as a formalization of the doctrine of "truth functions." One understands by truth functions extensional functions the arguments and values of which are propositions, e.g., $p \vee q$ (logical sum, disjunction), $p \supset q, \sim p, p \equiv q$ ⟦added above: pq logical product, conjunction, $p|q$ at least one false⟧. By extensionality, each such function can be represented through a table that specifies what truth value it takes for whatever distribution of truth values w, f of its arguments, for example, $\vee$ $\equiv$ Frege was the first one to make these clear.

Each meaningful formula of the propositional calculus is obtained through finitely iterated application of these truth functions on propositional variables. (Truth functions are operations in the sense of von Neumann, constants and abstractions don't find place here.) Two formulas are called equivalent if they result in the same truth value for arbitrary values of the arguments. Such equivalences can be used for the definition of single truth functions from others, e.g., $p \supset q$ by $\sim p \vee q$. In particular, *all truth functions* are definable in this way from each of the following three pairs: $\sim$ and $\vee$ (Post), as well as from $\sim$. and from $\sim$ $\rightarrow$, on the contrary from none of the single functions listed above (Hilbert-Ackermann). $\vee$ can be

3

24-423

defined from just $\rightarrow$, by $(p \rightarrow q) \rightarrow q$. $\vee$, on the other hand, is not definable from $\rightarrow$.[1] Sheffer was the first one to show that all truth functions can be obtained from one, namely from $p|q$, p is incompatible with q, i. e., $\sim p \vee \sim q$, as well as from $\sim p \,.\, \sim q$. These are even the only two with this property (Zylinski [ ] has investigated all the 16 truth functions with two arguments in respect of the definability of others from them.)

A formula that gives, for an arbitrary substitution for its arguments, always w is called generally valid (Hilbert-Ackermann) or tautological. This can be always ascertained through finitely many trials. The following interpretation of the calculus offers an elementary decision procedure for this: Let natural numbers be taken for the propositional variables, for $p \vee q$ the arithmetic product, for $\sim p$ $p+1$. Then it is sufficient to apply the rules of calculation *mod* 2 (a tautology results identically 0), Herbrand, Gegalkine.

Another procedure depends on the possibility to bring each formula

---

[1] ⟦$\rightarrow$ and $\vee$ should be changed, as indicated in a cancelled remark: but not the other way around⟧

into conjunctive normal form, i.e., the ability to give an "equivalent" formula that is a conjunction of disjunctions, where each disjunct is either a variable or a negation of a variable. There exists though, dually, always a disjunctive normal form, for the decision whether an expression is "satisfiable," i.e., whether there exists somehow a combination of truth values for which it becomes true.

One obtains from a generally valid equivalence, on both sides of which there occur only $\sim$, $\vee$, and ., again one such by exchanging $\vee$ and . (theorem of duality, cf.). (With an implication, one must exchange both sides simultaneously.)

To formalize the calculus so far described in a contentful way, one has still to indicate axioms and rules of inference, such that the generally valid formulas, and only those, are derivable from the axioms by the rules of inference. The axiom system is called complete in this case. Because of the mutual dependence of the truth functions of each other, it is sufficient to use one of these (or a pair) as a basic concept from which the others can be defined. Hereby the concept of a meaningful proposition is correspondingly limited. Other functions can then possibly be introduced as abbreviations. The first such system stems from Frege [ ],

4

24-424

Begriffsschrift. He uses $\sim$ and $\rightarrow$ as basic concepts and puts up six axioms, of which the third is superfluous. The rules of inference are (and the same in all the systems that follow, if the contrary is not expressly mentioned), the rule of substitution (arbitrary meaningful expressions can substituted for propositional variables in formulas already proved), and the implication rule, .

Other systems stem from Peano [ ], ($\supset$, ., $\sim$, 11 axioms); the rules of inference are not yet explicitly formulated, Russell [ ] ($\sim$, $\vee$, five axioms, the last one superfluous by Bernays, cf. p.), Hilbert [Ann 88].

The simplest system is the one of Lukasiewicz [ ] with $\sim$, $\rightarrow$ as basic concepts and the following three independent axioms ⟦addition to text indicated but not given⟧. Nicod is the first one to show that one can get along with a single (very complicated) axiom [ ]. ($p|q$ is his sole basic concept). The axiom of Nicod contains certain subformulas that are themselves tautologies, it is not "organic" in the terminology of Lukasiewicz and Wajsberg. Wajsberg gives in [ ] an organic axiom for the basic concept $p|q$. One can give also for the system with $\sim, \rightarrow$ as basic concepts a single complicated axiom, as shown by Tarski.[2] One can show, on the other hand, that there exist for each natural number n complete axiom systems of the propositional calculus with exactly n independent axioms, further, that there has to occur in each complete system at least one axiom with three different variables [proof not yet published].

To go over from one axiom system to another with other basic concepts, it is not sufficient to express the old axioms in terms of the new basic concepts, but one has to generally add certain new axioms (cf. Post).

Herbrand [ ] gives a system with the law of excluded middle as the only axiom and $\overset{?}{n}$ rules of passage. One shows, after Post, that all tautologies are provable from the system mentioned, by: 1.) showing that each formula of the shape $A \equiv N$ can be proved, where A is an arbitrary expression and N its conjunctive normal form, and 2.) that each tautological normal formula is provable.

5

24-425

Other proof methods stem from Lukasiewicz.

The propositional calculus is complete also in the sense defined on p. , i.e., the addition of an arbitrary non-derivable formula to the axioms makes *each* formula derivable [Post].

The proof of independence of the axioms succeeds through a small modification of the method of models that is common for the axioms of geometry, one that was given by Bernays [ ] and Lukasiewicz. One specifies for a certain system $\mathfrak{A}$ of elements (usually finite) two

---

[2] ⟦What follows is cancelled and written more compactly on the facing right page, then very lightly cancelled. The first version says that the axioms are independent, added in my reading.⟧

operations $\sim a$, $a \rightarrow b$ (in case $\sim$, $\rightarrow$ are the basic concepts) that produce elements of $\mathfrak{A}$ from elements of $\mathfrak{A}$. Further, a subset $\mathfrak{B}$ of $\mathfrak{A}$ is specified (the set of "distinguished elements" of $\mathfrak{A}$), namely so that if $x \supset y \varepsilon \mathfrak{B}$ and $x \varepsilon \mathfrak{B}$, then also $y \varepsilon \mathfrak{B}$. The two sets $\mathfrak{A}$, $\mathfrak{B}$ together with the operations $\sim$, $\rightarrow$ make up for what is known as a "logical matrix." If a formula A, under an arbitrary substitution of its variables by elements from $\mathfrak{A}$, always gives elements in $\mathfrak{B}$, one says that it fulfils this matrix. To prove the independence[3] of an axiom A, one has to specify a matrix for which all the axioms up to A are satisfied. The proof of freedom from contradiction of the above system can be carried through in an analogous way, through the matrix $\mathfrak{A} = \{w\,f\}, \mathfrak{B} = \{w\}$ with the meaning given above to $\rightarrow \sim$, etc. Not all proofs of independence can be carried through by finite matrices, a counterexample, cf. .

General investigations on the propositional calculus were conducted at the Warsaw seminar on mathematical logic and the results are published, without proof, by Tarski and Lukasiewicz. The authors consider these as a specialization of the general theory of formal systems presented above [ ], namely by interpreting S as the set of meaningful formulas that are built up from $\sim$, $\supset$ and variables and $F(X)$ as the set of formulas derived from the formulas of X through iterated substitution and the rule of implication. The axioms specified on p. [ ] are then satisfied. The concept formations announced there become possible. The set of tautologies is, in this sense, only one of infinitely many possible "systems" that one can obtain either through a "basis " (axiom system) or through a matrix. In the latter case, those and only those formulas are counted within the system as satisfy the matrix. Each system defined through a basis can be also defined

6

24-426

through a matrix, but not always through a finite matrix (even if the basis is finite). The intuitionistic system of propositional calculus

[3] This term, introduced in [ ], refers to quadratic tables through which one can represent the operation $\rightarrow$ for finite $\mathfrak{A}$.

would be a counterexample (Gödel).

The system of classical propositional calculus discussed above is determined by the matrix for w, f (two-valued logic), and one can consider certain other systems analogously determined by matrices as many-valued logics. Here the starting point is that the propositions can assume, instead or next to "true" and "false," even other values, e.g., necessary, possible, uncertain, known, etc. To be able to handle concepts of this kind (modal-logical concepts, especially that of necessity), Lukasiewicz gives the system of three-valued logic [ ], the three values of which one can interpret approximately as true, false, and undetermined.

In [ ], an n-valued system L_n of the propositional calculus is defined, through the matrices

$$\mathfrak{A} = \{0, \tfrac{1}{n-1}, \tfrac{2}{n-1}, \ldots 1\},\ \mathfrak{B} = \{1\},$$
$$\sim p = 1 - p,\ p \supset q = min(1, 1 - x + y)$$

Analogously the infinitely many-valued one when $\mathfrak{A}$ is the set of rational numbers $L_{\aleph_0}$. L_2 is the usual propositional calculus (we have for each n that $L_n \subset L_2$). Each L_m is axiomatizable (cf. p.) (this is undecidable for $L_{\aleph_0}$). Wajsberg gives an axiom system for L_3 with four independent axioms, together with a proof that all formulas of L_3 follow from them. In the case that $n-1$ is a divisor of $m-1$, we have $L_n \subset L_m$. In case, and only in case $n-1$ is a prime number, we have $\gamma(L_n) = 3$ (cf. p.).

Another interesting result from [ ] is that there exist, for each L_n and each

7

24-427

m, axiom systems with exactly n independent axioms, and that there must occur in each axiom system for the L_n an axiom with at least three different variables. Further, it will be determined of some formulas whether they are indecomposable, i.e., whether the set of consequences is also a set of consequences of two or more independent propositions. Other general investigations about the calculus of propositions were conducted already before the Poles by Post in

[ ]. He considered systems with arbitrary (possibly k-place) basic concepts, but also arbitrary rules of inference. Such a system S_1 is said to be representable in another S_2, if one can define functions from the ground functions of S_2, such that the same formulas are provable in S_2 about them as about the ground functions of S_1 in S_1.

Conditions will be given first of all for when a system is a "truth system," i.e., representable in the usual propositional calculus (i.e., in modern parlance, when it can be produced through a matrix with two elements).

Post gave already a many-valued propositional calculus T_n ⟦cancelled: with n truth values⟧, different from the ones discussed earlier, through the definition

$$\mathfrak{A} = \{1\,2\ldots m\},\ \mathfrak{B} = \{1,2,\ldots u\}$$

in which $u < m$, $\sim p \equiv p + 1(m), p \vee q = max(p\,q)$. He proved that each function of the m arguments from $\mathfrak{A}$ can be obtained through iterated application of these two.[4] He also sketched a method for obtaining finite axiom systems for his many-valued systems and for proving their completeness. The question about the conditions that the numbers m_1, u_1, m_2, u_2 must satisfy for $T^{u_1}_{m_1}$ to be representable in $T^{u_2}_{m_2}$ was solved only in part. Finally, he gave an interpretation to his many-valued logic by taking the variables to be ordered n-tuples of propositions [namely ones in which a false proposition is always followed by a false proposition] and the truth value of such an n-tuple to be the number of true propositions contained in it.

8

24-428

If among the tautologies of the propositional calculus only those are considered that contain just some functions not sufficient for interpreting the whole language (for example, only $\rightarrow$), there arises the problem of finding an axiom system for this narrower domain of formulas, one that also belongs to this domain. This problem was solved by Tarski [ ] for formulas that contain only the implication

---

[4] ⟦The sentence seems to have been changed halfway, with words that don't match well together.⟧

sign, or the restricted propositional calculus in the terminology of Tarski. It was solved by Lukasiewicz for formulas that contain only the equivalence sign. Wajsberg gives in [ ] necessary and sufficient conditions that tautologies with negations in them must fulfil for them to yield, when added to the restricted propositional calculus, an axiom system for the whole calculus. This result is partly extended to many-valued logic.

Hertz investigates an even narrower propositional calculus, by considering only formulas of the form $a_1\, a_2 \ldots a_n \rightarrow b$ (here a_i and b are propositional variables). The following are used as rules of inference:

1. From $b_1\, b_2 \ldots b_n \rightarrow b$ and $a \rightarrow b_i$ (for each i) follows $a \rightarrow b$.[5]

2. Each formula of the form $a \rightarrow a$ can be used.

3. Arbitrary variables can be added in the antecedent of an implication (no rule of substitution).

Hertz investigates especially the question when a system of propositions has just one basis and derives for it a necessary and sufficient criterion. He investigates further bases with the smallest ⟦sentence seems incomplete⟧

9

24-429

Russell already introduced, in [ ] and [ ], the all-operation $(p) f(p)$ (i.e., for all propositions holds) into the propositional calculus (extended propositional calculus). The concept "meaningful formula" is extended hereby, because quantification (p) can now be applied on propositional variables in arbitrary alternation, next to the previous operations.

The concept of "tautology" is definable analogously to earlier and remains even here decidable, because the question is with (p) just about the two cases of a true and a false proposition. Negation can be defined here from only implication, in the following way: $\sim p = (q)\, p \supset q$. Therefore $\rightarrow$ suffices as a sole basic concept.

---

[5] ⟦The qualification "generalized form of the syllogism" is cancelled.⟧

If one adds to the previous rules of inference the deletion and addition of the all-operator (cf. about it), each axiom system for the restricted propositional calculus is at the same time one for the extended one, as Tarski has shown.

One obtains an even more comprehensive system, called protothetic by Lezniewski, if one introduces variables f, g for truth functions, together with the accompanying all-operator, and possible further variables for functions the arguments of which are truth functions, and so on. Here it is sufficient to take equivalence $\equiv$ as the sole basic concept. By [ ], one defines the logical product from it, through $pq = (f).p \equiv .f(p) \equiv f(q)$, and then $p \supset q = .p \equiv pq$. It is essential for the definition of pq that f runs through the extensional functions, i.e., truth functions. Tarski gives, though, in [ ] also a (complicated) definition that is free of this restriction. A foundation for protothetic with just one axiom and

10

24-430

rule of inference was given in [ ]. One can even build an analogous system of protothetic for the many-valued propositional calculus.

Lewis made an objection to the concept of implication as defined through $\sim p \vee q$, namely that it does not convey that which is meant by "logical consequence." For it would not be correct, say, that from each false proposition logically follows each proposition, that from each proposition logically follows each correct proposition, and that out of two propositions, one always follows from the other. (Whereas the formulas $p \supset .q \supset p$, $\sim p \supset .p \supset q$, and $p \supset q .\vee. q \supset p$ hold for implication.)

He puts therefore up an axiom system with another concept of implication (strict implication) for which these paradoxical results are not derivable. This system has the basic concepts $-p$ (not), $p.q$ (and), $\sim p$ (impossible, i.e., false or contradictory on logical grounds) from which is defined: material (i.e., the usual) implication $-(p.-q) = p \supset q$, strict implication through $p \prec q = \sim(p-q)$, strict equivalence through $Mp = -\sim p$, necessity through $Np = \sim - p$, compatibility through ⟦missing⟧

Lewis [ ] had originally taken among the axioms also the following formula: $p \prec q \,.\!=\!.\, \sim q \prec \sim p$. It had, together with the other axioms, the consequence that $\sim p = -p$ became provable and therefore the calculus equivalent to the usual one (Post). Therefore the above axiom was weakened in [ ] into $p \prec q \,.\!\prec\!.\, \sim q \prec \sim p$. Parry was able to prove for this system, through the matrix method, that the above paradoxical propositions are not derivable for $\prec$. We have instead that each proposition strictly implies each necessary proposition and each impossible proposition each proposition, i.e., $Np \prec .q \prec p$ and $\sim p \prec .p \prec q$ are derivable. Even several non-paradoxical propositions become underivable, such as for example $p \prec (q \prec r) \prec q \prec (p \prec r)$. All the tautologies of the two-valued propositional calculus are, on the contrary, provable for the "material concepts" $\supset, ., \sim$, (even their necessity), so that this is contained entirely in it.

The Lewisian calculus allows to express, next to p and $-p$, the traditional modalities $Np, \sim p, Mp, -Np$. One cannot, though, show each modality (i.e., a sequence of the signs $\sim$) to be equivalent to one of these six.

11

24-431

This succeeds, as Becker shows in [ ], through the supplementary axiom $Mp \prec \sim\sim p$ (1). A weaker axiom, with more irreducible modalities, would be $Np \prec NNp$ (2) (Becker). The first axiom comes close to the above interpretation ($\sim p = p$ logically impossible (contradictory)), because it is equivalent to $NMp \vee \sim Mp$ and states therefore that a statement that claims the logical possibility of a proposition is either necessary or impossible, i.e., not true or false on empirical grounds. The second supplementary axiom is adjusted to the interpretation of not possible through "provably false" for which the first one is not plausible. A closer investigation of both axioms is in Parry who shows, among others, the the latter axiom follows from each proposition of the form NNq, by the Lewisian axioms.

A modality is called positive if it consists of an even number of the signs $\sim$ and $-$, otherwise negative. Becker shows that from the even weaker axiom $p \supset N^i Mp$ (for each i), and some propositions provable in Lewis, follows that all of the positive (and similarly all of the negative) modalities can be ordered through the relation $\prec$ into a series that is isomorphic to pairs of rational numbers.

Wajsberg [ ] has recently given for the axiom of Lewis, complemented by Becker's supplementary axiom, the following interpretation:

The variables denote all of the subsets of a denumerable set M.

$p \, . \, q$ denotes the intersection of the sets p and q, $\sim p$ the complement of p in M.

$\sim p = M$ if ⟦empty space⟧ $\sim p = \Lambda$ if

He shows that those and only those propositions are provable that always give the value M for arbitrary values of the variables under this interpretation. One finds newer investigations on the Lewisian calculus in [ ].

MacColl had already made an approach towards a calculus of modalities. Orloff puts up in [ ] a calculus that deviates from the Lewisian one, for the concept of "logically follows."

Modal logic has a certain formal relation to the system of intuitionistic propositional logic discussed on p. , because if one interprets the intuitionistic basic concepts in, e.g., the following way,

$$\neg p = -Np, \; p \supset q = Np \supset Nq,$$

$$p \vee q = -(Np - Nq), \; pq = p \, . \, q$$ [6]

then all the theorems of the system of Heyting become provable from the system of Lewis and the above two supplementary axioms, the formula $p \vee \neg p$ instead not (Gödel, to be published).

12

24-432

In the formal development of logic, the propositional calculus is followed by what is known as the narrower function calculus (Hilbert-

---

[6] ⟦changed from $pq = Np \, . \, Nq$⟧

Ackermann). Let whatever domain of things $\mathfrak{m}$ be given (domain of individuals), and whatever relations with arbitrary numbers of places (the number of places is indicated by upper indices, with one-place relations nothing else but classes):

$\varphi_1^1\ \varphi_2^1 \dots \varphi_k^1$ these are defined for the individuals of the domain

$\varphi_1^2\ \varphi_2^2 \dots \varphi_k^2$

By a proposition of the first level ⟦Stufe⟧ (counting proposition)[7] over these relations is meant a proposition that is built up from elementary components of the form $\varphi_1^1(x)$, $\varphi_1^2(x\,y)$, etc (here there can occur in place of the variables x, y also names for specific individuals a, b of the domain), through iterated application of the operations $\sim, \vee$, etc of the propositional calculus and of the quantifiers $(x), (\exists x)$. It is essential here that the all-sign refers only to individuals, never to classes or relations. The part of a formula to which a quantifier is applied in the build–up of formulas is called its scope. It is the only subformula that begins with (x) ⟦or $(\exists x)$⟧.

The proposition that the classes φ_1^1 and φ_2^1 are equivalent is, for example, no counting expression about φ_1^1, φ_2^1. The propositions of first order in the sense of p. [ ] are the counting propositions about the atomic functions that one assumed to be at hand in the world.

If in a counting proposition one replaces the letters φ_i^k that had earlier denoted specific relations through variables (and similarly the possible specific individuals $a, b \dots$), there arises a "meaningful expression of the narrower function calculus." Such an expression (F) is called satisfiable if there exist relations φ_i defined in whatever domain I (and possibly individuals a_i), such that if they are substituted in the place of the function variables (and the possible free individual variables) in F, there results a true statement. One calls such a system of φ_i, a_i also a realization of F. F is called generally valid in I if $\sim F$ is not satisfiable in I.

A formula that is satisfiable (generally valid) in I, is also satisfiable (generally valid) in each domain that is not of a smaller (not larger) cardinality. The converse does not hold and there exist, especially,

7 ⟧*Zählausdruck*, a term of Löwenheim's for first-order expressions.⟧

expressions that are satisfiable in each finite domain (see, e.g., Herbrand).

The famous Skolem-Löwenheim theorem holds, however, which states that a formula that is satisfiable in whichever domain, is already satisfiable in a denumerable domain.

13

24-433

This was proved first very laboriously in [ ], then more simply in [ ], in about the following way:

One can bring each formula, to start with, into a "normal form," i.e., give an equivalent one all the quantifiers of which stand at the beginning in a series (they form what is called a "prefix") and all of the remaining formula (that contains no quantifiers anymore, i.e., is what is called a "matrix") is contained in their scope.

Let it be assumed, now, that the normal form of a formula has the following specific form:

$$(x_1 x_2 \dots x_n)(\exists y_1 \dots y_m) M(x_1 \dots y_m) \tag{1}$$

Here M is a matrix that contains certain relation variables $\varphi_1 \dots \varphi_k$. (The general case can be handled similarly and was even led back in [ ] to the special case considered. We have to derive from the condition that (1) is satisfied in a domain I into its being satisfied in a denumerable domain. The condition states that there exist some relations $R_1 \dots R_k$ such that, for formulas that arise from M through substitution of the R_i for φ_i (let it be called M'), the following holds:

There exists for each n-tuple $a_1 \dots a_n$ of elements from I an m-tuple $b_1 \dots b_m$ in I such that $M(a_1 \dots a_n\, b_1 \dots b_m)$ holds.

One associates now, in virtue of the axiom of choice, to each n-tuple $\{a_i\}$ *a specific* m-tuple $\{b_i\}$, and let this be denoted by $b_i = f_i(a_1 \dots a_n)$. Then $M'(a_1 \dots a_n \dots)$ holds always, for arbitrary elements $a_i \, \varepsilon \, I$. When one now considers the denumerable domain I that arises through an arbitrarily iterated application of the function f_k to some $a_0 \, \varepsilon \, I$, then formula (1) holds already for this domain I' [about the elimination of the axiom of choice, cf. p.], by which the Skolemian theorem is proved for the special case concerned.

The Skolemian theorem can be extended to propositions that consist of denumerably many counting propositions built (in arbitrary iteration) through conjunction and disjunction, further to propositions in which there occurs the operation of forming a Dedekind cut, and to ones that contain denumerably many E signs in the prefix [ ].

A formula is called generally valid (tautological) if it is generally valid in each domain of individuals, and it is naturally required of a formal system for the eFK ⟦narrower function calculus⟧ that all and only those formulas that are tautological be provable in it. The Russellian axioms 1– indicated on p. 25 give one such, where as the rules of inference there has to be:

1. The rule of implication.

14

24-434

2. The rule of substitution: Arbitrary meaningful expressions can be substituted for function variables in propositional functions (with the limitation indicated on p.).

3. Free individual variables can be transformed into bound ones through an all-sign drawn on a whole formula.

Rule 3 and axioms . . can be replaced by a single rule 4 that reads as follows: One can infer from $A \supset B$ to $A \supset (x)B(x)$ and the other way around provided that the variable x does not occur free in A.

An essentially different formulation that connects to the "second method of Russell" indicated on p. . . was given in Herbrand [ ], and in a somewhat different form in [ ]. $\sim, \vee, (x), (\exists x)$ are the basic concepts ($P, \Phi x$, etc denote in what follows arbitrary expressions built from them). All formulas that arise from an arbitrary identity of the propositional calculus through the substitution of formulas *without* bound variables are axioms. The rules of inference are: 1. The above rule 3. 2. From $\Phi(xx)$ follows $(\exists y)\Phi(xy)$, i.e., a free variable can be transformed into a bound one at all or some places in which it stands, through an $\exists$ sign drawn on the whole formula. 3. $P \vee P$ inside a formula can be replaced by P. 4. The same with $\sim (t)\Phi(t)$ replaced by $(\exists t) \sim \Phi\, t$, and the other way around. ⟦5.⟧ $(t)\ R \vee \varphi(t)$ through[8] (Remarkably, the implication rule is not needed here.)

15

24-435

Herbrand extends the narrower function calculus through the introduction of "descriptive functions" and variables $(f, g, \ldots)$ that relate to them, the arguments and values of which are individuals (e.g., $a + b$ for the domain of natural numbers). The concept of a meaningful proposition gets extended correspondingly, when not variables but expressions built from f, g and variables can occur in the argument places of relations. The axioms and rules of inference remain the same, only 5 is changed so that there can occur instead of x an arbitrary expression that relates to an individual. The theory is not in principle extended, because one can always replace functions by relations, such as for example $x + y$ by the three-place relation $z = x + y$, cf. also p. . It is, though, more convenient in the Herbrandian formulation.

Identity (between individuals) is also sometimes counted among the basic concepts of the narrower function calculus, with $x = x$ and $x = y \rightarrow \varphi x = \varphi y$ added as axioms. The theorem by which a formula that is satisfiable in a domain I is also satisfiable in each more comporehensive domain ceases to hold.

It is very easy to show the freedom from contradiction of the above system (cf.). (The same method suffices also for the Russellian system together with choice and reducibility, but without the axiom of infinity.)

A proof for the axioms of Russell's system was sketched in .

The completeness of the above axiom system, in the sense that generally valid formulas are provable, was proved by *Gödel* in [ ] in the following manner: The claim to be proved is equivalent to having each meaningful formula of the calculus either satisfiable or re-

8 ⟦The phrase is incomplete. What is known as Herbrand's rule of passage is the inference from $(t)(R \vee \varphi(t))$ to $R \vee (t)\varphi(t)$, referred to as 5 on the next page.⟧

futable, where by "A is refutable" is meant: "$\sim A$ is provable." One can limit oneself in the proof to formulas of the shape

$$(x_1)\dots(x_n)(\exists y_1)\dots(\exists y_m)M(x_1\dots y_m) \tag{1}$$

because one can indicate, by Skolem, to *each* formula a formula of this shape that is satisfiable exactly when the original one is (and refutable, respectively).

Let us now assume that (1) is satisfiable, i.e., holds for some system of relations $R_1, R_2 \dots$ in a domain I. We denote by M' the formulas that arise through substitution of the R_i, and let $a_0 \,\varepsilon\, I$. It follows then, when one substitutes $a_0 \dots a_0$ for $x_1 \dots x_n$ in (1), that there exist $a_1 a_2 \dots a_m$ such that $M(a_0 a_0 \dots a_0\, a_1 \dots a_m)$ holds. It follows further similarly that there exist individuals $a_{m+1} \dots a_{2m}$ such that $M(a_1 a_0 \dots a_0\, a_{m+1} \dots a_{2m})$, etc, always substituting in order all the possible n-tuples of the a_i for $x_1 \dots x_n$. Let the logical product of the n first expressions thus built be called the n-th derivative ⟦Ableitung⟧. It contains no quantifiers, but just the operations

16

24-436

of the propositional calculus. If one of the derivatives is a contradiction of the propositional calculus, it is refutable in the above formal system, and then also the formula in (1). If no derivative is a contradiction, i.e., if all are satisfiable, one can show that there exists a domain of individuals and a system of relations for which (1) holds, i.e., (1) is satisfiable in this case, by which the claim that (1) is either satisfiable or refutable is proved.

The n-th derivative can be defined in an entirely analogous way for formulas the prefix of which does not have the above specific shape. Hereby the change of the Skolemian reduction mentioned above becomes superfluous.

The method used in the proof of completeness gives at the same time a proof of Skolem's theorem that is independent of the axiom of choice, as was already indicated in [ ], and carried through in [ ]. The above proof of completeness can be easily extended to formulas with = [ ], and the same for formulas with function variables. One obtains, after Wajsberg, a complete formalization for sets of formulas that are generally valid for a domain of individuals with exactly n elements, if one adds to the above axioms any formula that is generally valid for n but not $n+1$ elements.

The above proof is not finite (it uses the law of excluded middle for infinite totalities). What is proved finitely is that the specification of a formula that is generally valid but not provable from the above axioms would have as a consequence a contradiction with a (sufficiently wide) formal system of classical mathematics.

This question was treated by Herbrand in [ ] from a finitary standpoint, by which, however, the proof of the completeness theorem became impossible. One can, instead, show in the finitary way that the refutability of a formula is equivalent to the existence of a number for which the n-th derivative is a contradiction [ ]. Here the number n can be constructed from the formal refutation and the other way around. This theorem delivers a method for the construction of a finitary proof of freedom from contradiction [cf.].

24-437

If all of the axioms of an axiom system (cf. p.) are counting propositions about the basic concepts (first-order theory), one obtains the formalization of such a theory by adding to the above axioms and rules of inference of the narrower function calculus the specific axioms of the theory (their conjunction will be called A). The fact that a formula P is a consequence of the axioms A is equivalent to the general validity of $A \supset P$, and the completeness theorem guarantees, for the case of axiom systems of the first order, the equivalence of the two concepts of consequence (discussed on p.), as well as the theorem by which each such system either has a realization or contains a contradiction that can be produced by finitely many inferences.

The theorem of Skolem states that each system free from contradiction has already a denumerable realization. It is in principle possible to conceive of each axiom system as one of first order, by taking the set of basic objects as a special kind of individuals,

17

24-438

the relations of these to the other basic objects being regulated by special axioms of set theory. This conception is in particular the only possibility with the axiomatization of set theory, because one cannot very well use the axioms in a contentful way in the setting-up of the axioms (i.e., one must not count them within the "basic discipline"). Skolem's theorem is, then, in any case applicable here and it states: There exist for each axiom system of set theory that is free from contradiction realizations in which all of the sets are denumerable. This leads in no way to an antinomy, for the reason that even if one still distinguishes sets of different cardinalities in the axiomatized set theory, by that is meant that they cannot be mapped one on another through relations that occur within the theory (or within a realization of the theory), something that does not exclude the existence of such relations outside the system (set-theoretical relativism, cf. especially Skolem 1929).

Skolem expresses in [ ] the conjecture that this relativity can be extended to, e.g., the concept of well-ordering, in that one will be able, for some realizations, to specify sets that are well-ordered within the theory but have subsets with no first element outside the theory. He makes this plausible in an example of a simple axiom system. Skolem conjectures further that the concept of a "finite set" will turn out relative in this sense, something that has been confirmed by the result discussed on p. [1929 and about arithmetic].

Even the formal systems introduced in parts 2 and 3 (Hilbert, Russell) can be conceived of in a similar way as first-order formal theories, by taking, say, the predicative functions to be the individuals, and the relation that obtains between x and φx whenever φx holds as the basic relation. It must be noticed here that some axioms have to be conceived of as rules for the formation of infinitely many counting propositions (e.g., the axiom of reducibility). Further, there can appear different kinds (types) of individuals that must not, in such cases, be substituted one for another (e.g., the Russellian types of functions). Herbrand [ ] shows, however, how one can always specify an equivalent theory with just one type of individuals. It is even possible to always manage with finitely many axioms, through a reformulation, as was shown with the example of set theory in [ ].

The investigation of mathematical theories as axiom systems of first order was pursued especially by Skolem in [⟦1920⟧]. He was also the first one to attempt the construction of finitary proofs of freedom from contradiction from this point of view.

18

24-439

He proved in [ ], for example, a theorem that states in the terminology introduced on p. [ ] that if no derivative of a formula F is a contradiction, then this holds also for the formula $F \,.\, A$ when A is the law of excluded middle expressed for whatever relations that occur in F. (Freedom from contradiction is conserved under the addition of the law of excluded middle.)

Herbrand brought then this method to full clarity in [ ]. It is easy to see that an axiom system of the first order is free from contradiction exactly when the negation of the logical product P of the axioms is not generally valid, i.e., not derivable in the narrower function calculus (here the basic relations of the system are to be taken as the function variables). It follows by p. [ ] further that such an axiomatic system is free from contradiction if and only if no derivative is a contradiction, i.e., the n-th derivative has a realization for each n. *Whether there exists such a realization is, for each single n, a decidable question,* because no quantifiers occur anymore in the n-th derivative, (*and therefore one has just to determine of a certain formula of the propositional calculus if it is a tautology*). To prove the freedom from contradiction of the axiom system, instead, one must specify a procedure by which *each* derivative delivers a realization, i.e., one has to construct an infinite sequence of realizations (Herbrand calls such *Champ infini*). This is a decidable question for each single n, because for that it is sufficient just to decide of a certain formula of the propositional calculus if it is a tautology.

This construction succeeds always when one has a "finite realization" for the axioms themselves, i.e., one in which all the existential assertions that occur in the axioms can be confirmed through construction. Herbrand's proof shows therefore that if one applies the above partially transfinite rules of inference of the narrower function calculus[9] to finitely provable assertions, one does not arrive at any contradiction.

One case of application is the system of classical arithmetic discussed on p. , where one leaves out the function variables and axiom IV 2 that relates to them, and limits the use of the axiom of complete induction IV 4 to formulas $A(x)$ in which

19

24-440

ε and π don't occur, so ones that are built up from recursively defined functions with the help of only = and the operations of the propositional calculus. This proof of freedom from contradiction is carried through in [ ] in which ε and π, the symbolism of the function calculus ((x), $(\exists x)$, together with the pertinent rules of inference) are laid as a basis, and also the new rule of Hilbert, *Annalen*, is involved. This is the finite proof of freedom from contradiction known today, because the system of von Neumann (cf. p.) is, without axiom 4, much weaker (because no recursive definitions are possible), and the proof of Ackermann contains essential gaps and errors.[10]

The system of Herbrand for arithmetic is sufficient for the building up of classical number theory (excepting the analytical) and for most of algebra. One could probably prove by the same method the freedom from contradiction of the *Principia Mathematica*, together with infinity and choice, but without reducibility, as well as analysis to about the extent of Weyls' *Kontinuum*. The difficulties of principle discussed on p. [ ] stand in the way of a further extension of finitary proofs of freedom from contradiction.

---

[9] Namely, the law of excluded middle is, for example, provable from them in the form $(x)F(x) \vee (\exists y){\sim}F(y)$. ⟦The formula is heavily changed from something else; The missing negation has been added.⟧

[10] ⟦The following is cancelled but clearly readable: As concerns the axioms with function variables, certainly wrong (cf.), and also in the case that only induction axiom IV 4 is included for arbitrary formulas A, its feasibility is made extremely improbable.⟧

What is known as the decision problem is to be kept distinct from the completeness of the narrower function calculus discussed on [p.]. It requires the specification of a procedure that allows, for each formula of the narrower function calculus, to decide in finitely many steps whether it is generally valid (which is to say, by p. , essentially the same as): Is it provable from the axioms of the narrower function calculus ? A sharper formulation of the problem that stems from . . requires a more precise determination of the domains of individuals (i.e., individual domains with what cardinality) in which an arbitrarily presented formula is generally valid and in which not.

Because of [p. of the report], a solution of the decision problem would at the same time solve the problem whether an arbitrary axiom system of the first order (with finitely many axiom) is contradictory and whether whatever given specific proposition follows from it,

20

24-441

and thereby these problems would be solved for the formal systems of mathematics of parts III and IV. One can further show that the decision problem is equivalent to finding a procedure that allows to decide, of an arbitrarily defined recursive property of natural numbers, whether it belongs to all numbers [ ], something from which would result the solution of, e.g., the Fermat or Goldberg problems. A solution of the decision problem would have, by what was said above, as a consequence the solvability (at least in principle) of almost all mathematical problems that occur in practice, which amounts doubtless in the very most cases to the question whether a specific proposition is derivable from some specific axioms. However, the same difficulties of principle stand in the way of a solution.[11]

---

[11] ⟦Long cancelled passages each with a similar content follow, with the "utter improbability" of a solution of the decision problem mentioned four times, the last one clearly indicating the incompleteness results as a ground. The final statement is: The decision problem contains therefore in it practically all the problems that occur in mathematics. (Herbrand calls it "the fundamental problem of mathematics." The possibility of its solution is, however, utterly improbable.⟧

So far, certain special cases were cleared on the one hand, and on the other hand, reduction theorems were proved that led the solution of the general problem back to a corresponding one with a certain special structure.[12]

⟦1.⟧ In the following, we lay the dual formulation of the problem as a basis, i.e., the question of when a formula is satisfiable. We call two formulas of equal value if they are either both satisfiable or both not satisfiable.[13] M denotes always a matrix in which there occur no descriptive functions. It has been shown: One can give to each formula with the = sign one of equal value without the = sign [ ]. Analogously for formulas with and without function variables [ ].

2. One can give to each formula one of equal value of the shape $(x_1 \ldots x_n)(\exists y_1 \ldots \exists y_n)M(x_1 \ldots x_n)$

$20\frac{1}{2}$

24-442

[ ].

3. One can specify for each formula one of equal value with just two-place relation variables (binary counting expressions), Löwenheim, Herbrand, Kalmar; further one with just one three-place or one with just three two-place relation variables, Herbrand.

4. One can specify for each formula one of equal value with a prefix of the shape $(\exists x_1) \ldots (\exists x_s)(x_{s+1} \ldots,$ and a binary to a binary one, Kalmar ⟦1929⟧.

Of the special cases of the decision problem, the 1–10 given below have been cleared so far. Of these, 1–8 depend on the theorem by which the satisfiability of a formula is equivalent to the satisfiability of all of its derivatives and the fact that one can in certain cases readily continue realizations of the first derivative to the following

[12] ⟦Gödel rewrote his account of the decision problem from here on to page 443 on detached pages. The last third of this page is lightly cancelled, the next page has a regular diagonal grid of lines drawn over the whole page, with the syllable count kept. Page 443 has the first three lines cancelled.⟧

[13] Formulas equal in value clearly need not be equivalent.

ones.[14]

1.) Formulas of the shape $(\exists x_1 \dots x_n)(y_1 \dots y_m)M(x_1 \dots y_m)$, Herbrand.

2.) Formulas of the shape $(x_1)(\exists y_2)M(x_1, y_2)$, Bernays.

3.) Formulas of the shape $(x_1)(\exists y_1 \dots y_m)M$, Ackermann, Skolem, Herbrand,

and of the shape

4.) $(\exists x_1 \dots x_n)(y_1)(\exists y_2 \dots y_m)$, Ackermann, Herbrand.

5.) Conjunctions of finitely many formulas of the shape 3, in which propositions can also have an $\exists$ sign in front of the whole formula.

8.)[15] For formulas of the shape $(\exists x_1 \dots x_n)(y_1, y_2)M(x_1 \dots)$, (Gödel, so far only a sketch of proof published).

The theorem by which there exists a realization in a finite domain if there exists one at all, is proved for cases 1–6

$21\frac{1}{2}$

24-443L[16]

and one can estimate the number of its elements from above with the help of certain numbers given by formulas (m, n the numbers of relation variables that occur in them) [Bernays, Ackermann] (by which the problem is solved even in a strict sense).

⟦A red circle indicates an insertion that is found on a detached page not included in the microfilm edition of Gödel's papers.⟧

---

[14] M denotes always a formula without apparent variables and also without = signs and variables for descriptive functions.

[15] ⟦Originally 6. The following, originally added at right with numbers 7 and 8 changed into 6 and 7, are cancelled:

6.) For formulas with an arbitrary prefix, the matrix of which is a conjunction of elementary components or their negations.

7.) For formulas of the shape $(x_1 \dots x_n)(\exists y_1 \dots y_m)M(\quad)$, with at least m different argument symbols in relation variables in M throughout.⟧

[16] ⟦The first four lines of this page originally continued the preceding page, with additions at the facing page. They have been cancelled but are given here to maintain continuity of text.⟧

One can reduce the decision problem, to begin with, to the case in which no = signs and no variables for descriptive functions occur in the formula at hand, i.e., one can show that for *each* formula, a formula of equal value can be given that contains no equality sings and no variables for descriptive functions. Two formulas are here said to be of equal value if they are either both satisfiable or both not satisfiable (formulas of equal value need of course not be equivalent). The following holds further for formulas without = signs and without descriptive functions: One can indicate for each such formula:

1. One of equal value of the shape $(x)\dots(Ey)\,M$ (in which M contains no quantifiers anymore).

2. One of equal value with just two-place relation variables.

3. One of equal value with altogether just one, and precisely three-place relation variable, and:

4. One with altogether three precisely two-place function variables.

(5. One of equal value of the shape . .   .)

In cases 2, 3, 4, also the sharpened decision problem is reduced back to the corresponding one for formulas of equal value.

As concerns the special cases of the decision problem dealt with so far, we shall limit ourselves for now to the case of formulas without descriptive functions and without = signs. If such a formula contains in normal form only all-signs, or if all the E-signs are before the all-signs, it is easy to decide whether it is satisfiable or not [   ]. Bernays and Schönfinkel solved this problem for formulas of the shape $(x)(\exists y)F(xy)$, and its generalization, Bernays, Skolem, for ones of the shape $(x_1)(\exists y_1 \dots z_1)$. The procedure can be extended to formulas of the shape   , as well as conjunctions of finitely many such formulas. It turns out in all the cases discussed so far that if a formula is satisfiable at all, then already in a finite

24-379L

domain, the cardinal number of which can be delimited from above (Bernays, Ackermann) on the basis of numbers given by the formula (the number of relation variables that occur there, etc). Hereby the decision problem is solved in a sharp sense.

The most general case of this method suffices still for formulas of the shape ( ) (Gödel), further ⟦added above: and treats⟧ formulas of the shape $(x_1 \ldots x_n)(\exists y_1 \ldots \exists y_m)$ in which there occur in M relation variables with at least m different argument symbols throughout [ ], as well as formulas with an arbitrary prefix the matrix of which is, still, a conjunction of elementary components and their negations.

The solution of the decision problem depends in all the cases considered on the theorem by which the satisfiability of a formula is equivalent to the satisfiability of all of the derivatives, and on the fact that one can continue without further ado, under the circumstances, from the first derivative to the successive one (and admittedly usually because there appears a sort of periodicity).

Another method calls for both of the following cases: Ramsey solves the decision problem in the sharp sense for formulas of the shape

24-443R

$(x_1 \ldots x_n)M(x_1 \ldots x_n)$ in which also the sign = can occur in M, by showing that such a formula F is either satisfiable in each or in no domain, or that a number h exists such that it is satisfiable in one and only one domain with less than h elements. The upper bound $n!!!$ is given for h in case there occurs just one relation variable in M, namely a two-place one (otherwise an essentially higher bound).

24-443L ⟦continued⟧

Formulas in which there occur just one-place relation variables (classes) have the property that if they are at all satisfiable, then they are already satisfiable in a domain of 2^n individuals, with n the number of class variables that occur in the formula in question [Hilbert-Ackermann]. Herbrand and Gegalkine give an elegant proof, by showing that each formula in which there can occur the class variables $\varphi_1, \varphi_2, \ldots \varphi_n$ is equivalent to a truth function of the 2^n possible propositions of the form

$$(\exists x)\varphi_{i_1}(x) \,.\, \varphi_{i_2}(x) \ldots \varphi_{i_k}(x) \,.\, \overline{\varphi_{i_{k+1}}(x)} \ldots \overline{\varphi_{i_n}(x)}$$

Here i_r takes, for $r = 1$ to n, all natural numbers between 1 and n. These 2^n propositions can take, for φ_i correspondingly chosen, independently from each other all possible truth values, with the exception that they are all false.

Behmann extends this decision procedure to the case that even class variables are bound by quantifiers in arbitrary ways (higher function calculus) and that possibly also the sign = occurs. He proves that such a formula is correct in almost all[17] domains of individuals, or false in almost all, with possible exceptional cases that can always be effectively given. (More about this, cf. .)

It is possible to give, for several systems of axioms of first order, decision procedures for specifying whether an arbitrary counting proposition a about the basic concepts is derivable from them, and to prove their completeness in the sense that either a or $\sim a$ is always derivable. Such a system exists, for example, for the operation $+1$ (successor) as a basic concept and the natural numbers as the domain of individuals (Herbrand in), further for the relation $<$ in the individual domain of natural numbers, Langford *Annalen* 2, for the relation $<$ in the domain of fractions (dense order), Langford *Annalen* 2, Skolem,

$22\frac{1}{2}$

24-444

further, for the operation $a + b$ in the domain of individuals of whole numbers. There occurs in each of the systems an equality relation $x = y$ with the axioms that belong to it, further some basic constants. The latter system offers at the same a decision procedure for a certain category of number-theoretic propositions (namely for propositions about the concepts +, =), one that is, however, mathematically not of much interest, because such theorems can be used merely for expressing simple properties of linear congruence. Skolem gives in ⟦1930⟧ a decision procedure for a somewhat wider domain of sentences, by adding the operation of multiplication by constant rational numbers (only one factor is allowed to be a variable or an

[17] Domains of individuals with the same cardinal number are to be identified here.

expression that contains a variable, the other has to be a specific rational number) and the operation of forming the next-smallest whole number. He sketches further a method that is sufficient for the operation of multiplication and the relation of equality in the individual domain of the natural numbers. (If one could extend the decision procedures that one knows until so far for only addition and multiplication each, to sentences that contain addition as well as multiplication, one would (as results from [ ]) already master the whole of number theory (inclusive Goldbach, Fermat), the Riemann hypothesis, etc. However, difficulties of principle stand in the way of such [cf. p.].

The decision procedure, to the extent that one has such, depends in all cases on the possibility to specify, for each counting proposition function with the basic concepts in question, an equivalent matrix, or in other words, there exists to each matrix $M(x, \dots y)$ an $M'(y\,z\dots)$ such that $(\exists x)M(xyz\dots) \equiv M'(yz\dots)$ holds and is provable from the axioms. One can eliminate by this successively the apparent variables and obtains in the end an equivalent proposition that is easy to decide, one that contains no variables but just special elements of the domain of individuals (numbers).

There are several systems for which decision procedures have been successful for general propositions, i.e., ones that don't contain any existence signs. Langford ⟦1927⟧ presents such a system for the betweenness relation, Skolem ⟦1920⟧ for a part of the calculus of classes and for the incidence axioms of the plane.

$23\frac{3}{4}$

24-445

One can represent the counting expressions over one-place relations (classes) in another symbolism, through the operations $a + b$ (union class of a and b), $a\,.\,b$ (intersection class), $\overline{a}$ (complement class), $=$ (identity between classes). An expression that is built up from whatever variables by the said operations is called a "logical function of the variables." Each counting proposition about $a_1 \dots a_n$ is equivalent to a truth function of equations between logical functions of the a_i (this results immediately from, e.g., the Herbrand normal form mentioned on p.). The theory built upon the basic concepts $+, ., ^{-}, =$ can also be developed as an independent contentful axiomatic theory. It appears in this form as the first part of logic that has been treated mathematically.

The first beginnings go back to Leibniz who already formulated a series of propositions that he deduced from certain axioms [cf. especially]. Boole was then the first one to develop a comprehensive theory in which, however, the analogies to arithmetic were pushed all too strongly into the foreground, to the detriment of conceptual clarity. The theory was freed from this defect by Jevons and Peirce who at the same time enriched it with new theorems and methods. Further results stem from Schröder who gave an extensive and comprehensive presentation in [ ].

The Aristotelian syllogistic appeared as a minute part of the theory. All of the valid modes were combined into a single equation by Ladd-Franklin in , namely . (What does the empty class mean ?)

Jevons [ ] gave already a system of axioms for the above concepts, and then Whitehead , in analogy to the usual arithmetic, in a form that is easy to survey. Huntington proved that some of Whitehead's axioms are superfluous and put up the following independent system with the basic concepts +, . =:

$a + b = b + a$, $ab = ba$, existence of two classes 0 (the empty class) and 1 (the universal class), so that for arbitrary $a, a + 0 = a$, $a \,.\, 1 = a$.

The formulas $a(b + c) = ab + ac$, $a + bc = (a + b)(a + c)$ are the two distributive laws, of which the second (addition in a product) has no analogy in arithmetic.

$24\frac{3}{4}$

24-446

Finally, there exists for each a at least one $\overline{a}$ such that $a + \overline{a} = 1$, $a \,.\, \overline{a} = 0$. Huntington gives also a system in which +, =, and Lewis one in which ., =, are the only basic signs, but complicated axioms because of that. Huntington gives in [ ] an especially simple system with the basic concepts $+, \overline{\ \ }, =$.

Schröder (resting on Peirce) gives a foundation for the calculus of classes in a different way, by using the inclusion relation $a \subseteqq b$ (subsumption) between two classes as the sole basic concept. He defines $a = b$ from it as $a \subseteqq . b \subseteqq a$, and $a + b$ as the class that is contained in all classes in which both a and b are contained (the product analogously), further 0 as the class that is contained in all classes (1 analogously), and the complement of a as the class b for which $a + b = 1$, $a \cdot b = 0$. The existence (not uniqueness) of $a + b, ab, 0, 1, \bar{a}$ must be required through special postulates, ones Schröder tried to avoid by introducing them in a manner of speaking as "ideal elements" in case they were not already at hand among the original classes. This attempt, though, was not entirely successful. Schröder requires for $\subseteqq$ merely reflexivity and transitivity, and the second direction of the distributive law, namely $a(b + c) \subseteqq ab + ac$, must be postulated separately. It is easy to prove the independence of the second distributive law from the rest of the postulates, cf. further Peano in . Huntington shows the independence of the other axioms of Schröder.

$25\frac{1}{2}$

24-447

Schröder investigates under the name calculus of groups the system in which the second distributive law and the one about the complement are left out. A system equivalent to this, with the basic concepts $+, .$, are the field axioms of Menger that Menger lays as a basis for projective geometry ($a + b =$ connecting space and $a \,.\, b =$ intersection space of two linear spaces a, b). Wernick puts up in [ ] a formal system with precise rules of inference for the subsumptions between two logical functions that are derivable without the second half of the distributive law. He gives also a decision procedure for when such a formula is provable in his system (though only for those that contain no negation). If one adds the second distributive law to the axioms of Wernick, all the subsumptions that are provable in the usual calculus of groups become derivable.

Huntington replaces the second distributive law through an axiom that is simpler to grasp (cf. in this respect also Neumann and E. Müller), namely: ⟦not shown⟧. He conjectures that it is equivalent to the uniqueness of complement formation. This is, however, challenged by Berg⟦mann⟧ [ ], for a wider concept of complement formation. Namely, the complement x for each two things a, b (for which $a \subseteqq b$) is defined through $a + x = b$, $ax = 0$.

Certain subsystems of the usual calculus of classes are investigated also in Klein?, especially in relation to the question when there exists a unique decomposition of the elements in prime elements.

26

24-448

Sheffer gave for the calculus of classes a system with $a|b$, i.e., $\overline{a}\overline{b}$, as the sole basic concept, with four independent axioms (Bernstein diminished in [ ] the number of axioms by one, and Taylor achieved by a further modification complete independence). Yule used $ab = 0$ as the sole basic concept. All of the systems mentioned can be shown equivalent. Of their realisations (called logical fields by Huntington), only those with finitely many elements are known so far. Each such is isomorphic to the subset of a finite set M with $+, ., ^{-}$ to be understood as sum, intersection, and complement in M. It then follows that the number of elements of a logical field must be a power 2^a (Huntington). As concerns more detailed investigations in this direction ⟦cancelled: Tagaki⟧ cf. ⟦Tagaki (1931)⟧. There exist in infinite realizations no "individuals," i.e., elements without proper parts, cf. Schröder.

If in a correct sentence of the calculus of classes, expressed in the symbols $+, ., ^{-}, =, ⋹$, one exchanges + with . and $\subseteqq$ with $\supseteqq$, there arises again a correct sentence (duality principle). It was stated already by and derived by Schröder from the fact, easy to see, that it holds for the axioms. Another derivation follows from the fact that there exists a one-to-one mapping of the domain of classes to itself, by which + is transformed into . , $⋹$ into $⋺$, and $^{-}$ as well as = into themselves, namely the mapping that transforms each class into a complement. [This conception lies at the basis of, for example, [Neurath], though not in a concrete form.] Hereby the same theorems hold for the pair of functions $+,.$ as for $.,+$. There exist, though, still infinitely many pairs of logical functions for

27

24-449

which we have, e.g., ⟦space left⟧. All such possible function pairs were ascertained by . There belongs to each such a one-to-one mapping of the domain of classes to itself, conveyed through logical functions. Functions and relations that go over to themselves under all one-to-one mappings, conveyed through logical functions, are called invariant. Necessary and sufficient conditions for invariance are specified in [ ] and [ ], and an axiom system with the invariant basic operation $a'b' + a'c' + b'c'$ is put up.

Kempe had laid even earlier the invariant triple relation $ab \supseteqq c \supseteqq a + b$ (logical between) as a foundation of a system. It shows remarkable analogies with the geometrical between, so that Kempe added to his axioms for the logical between just one axiom to arrive at an axiom system for affine geometry. He claims further that certain subsets of the basic objects of his logical system are directly isomorphic with the affine space.

The system of Kempe is not flawless, because of a maximal axiom that is formulated in an unclear way. It was essentially improved by Royce who also succeeded in dealing with sums and intersections of infinitely many sets and to derive the axiom of continuity within the geometric interpretation, through a generalization of the betweenness relation of Kempe. Royce sketched further a proof for the above claim of Kempe about the relation to geometry and conjectured that one will be able to represent even mathematical systems other than geometry in the same way, within a logical system (through the specification of subsystems isomorphic to them). By this he foresaw a new fruitful method for the derivation of mathematics from logic, quite different from the one of Frege, Peano, and Russell, that so far, however, has presented no further results.

28

24-450

At the center of the specific problems of the calculus of classes stand what are known as the elimination problem and the resolution problem. The first consists in the following:[18] Let $f(a,b\ldots k)$ be an expression built up with the help of $+,.,\,^{-},=,\underline{\in}$, and the operations of the propositional calculus ($\vee$, $\sim$ etc). Necessary and sufficient conditions are to be given that $b,c\ldots k$ must fulfil for there to exist a class a for which $f(a,b\ldots k)$ holds. Here it is required that the condition be given again in the form of an expression $g(b\ldots k)$ of the kind characterised above. $g(b\ldots k)$ is called the resultant of f in relation to a. The problem is easily reduced back to the case that $f(a,b\ldots c)$ is a conjunction of equations and subsumptions. Namely, each such is within the calculus of classes equivalent to a single equation of the form $f(x,b\ldots k)=0$, and these can again be written in the form $Ax+B\overline{x}=0$ in which A,B depend only on $b\ldots k$. We have, namely, for arbitrary logical functions $f(x)=f(0)x+f(1)\overline{x}$ (analogously for functions with arbitrarily many variables). The resultant of the above function reads $AB=0$. It is equally easy to build the case of a conjunction between an equation and an inequality [ ].

One has to rather (as Behmann shows in) add two new concepts, namely "class k contains at least n elements" and "class k contains at most n elements" (for each arbitrary natural number n). It corresponds, in the representation of the calculus of classes in the symbolism of the function calculus, to the addition of the identity between individuals among the basic concepts. [For example, one can express that k contains at least two elements like this: $(Exy)\,x\neq y\,.\,k(x)\,.\,k(y)$.]

29

24-451

It is possible, for each expression built up from the earlier and the two new basic concepts, to represent the resultant through an expression that also contains only the mentioned concepts (as shown first by Behmann in). This delivers at the same time a solution to

---

[18] ⟦Gödel has started to change some symbols, f to R, g to S, and a to x, soon abandoned. I have kept the original symbols.⟧

the decision problem for the calculus of classes, in the form indicated on p. , because the bound class variables of an expression can all be removed through the successive formation of resultants.

The resolution problem consists in finding, for a logical equation $f(x_1 \ldots x_n) = 0$ that can possibly contain also undetermined coefficients a_i, expressions for the unknowns

$$x_1 = g_1(u_1 \ldots u_m)\; x_2 = g_2(u_1 \ldots u_m) \ldots x_n = g_n(u_1 \ldots u_m)$$

such that if the u_i run independently through all of the classes, the g_i deliver all the solutions of the given equation and nothing else.

The solution for the general equation $Ax + B\overline{x} = 0$ reads $x =$. Because of the above solution of the elimination problem, it is possible to reduce back the resolution of an equation with several unknowns into an equation with one unknown. Hereby one obtains a solution with as many parameters as unknowns. (It is impossible to succeed with less in the general case, cf. Löwenheim). The solutions thus obtained exhibit, however, an unfounded lack of symmetry in the unknowns. Also, it is not possible to recognise the law of formation of the general case of n unknowns. Whitehead gave more symmetric solutions (after a method of Johnson), for which $2^n - 1$ parameters were required for n unknowns.

Löwenheim gave in [ ] a surveyable law of formation for certain particular cases from which the general one can be obtained, in a work rich with mathematically interesting methods. The problem of symmetric solutions was posed by Schröder in a precise form, namely, he required that each permutation of the x_i and their elements and the undetermined coefficients that possibly occur in the given equations that transforms the equation into itself should also transform the solution to itself when one adds still one suitable permutation of the u_i and their elements. Schröder gave in [ ] a whole number of examples of symmetric

30

24-452

solutions, among them also for the general equations with two unknowns, further an example for the impossibility of a symmetric solution, cf. .

If in the above formulation one leaves with x_i and u_i out the words? and their negations, one obtains a different symmetry property. Skolem gave in [ ] necessary and sufficient conditions for the symmetric solvability of an equation in this sense. It emerges from these, among others, that the general equation with two or more unknowns is not such.

A particular kind of problem is concerned with the finding of all formulas of a determinate kind and with indicating their number. For example, the number of propositions that one can make about n classes with the help of the symbols of the calculus of classes and the propositional calculus, without the use of quantifiers, amounts to $2^{2^{2^n-1}}$ (Peano).

Two propositions are said to be of the same type if they arise one from the other through an exchange of the variables or their negations among themselves. There exist as many types as there are incongruent sets of an n-dimensional cube. Their number could be first determined for certain special cases, cf. Schröder, Clifford, Nagy. Poretsky determined all the equivalent formulas for which one obtains an equation $f(a_1 \dots a_n) = 0$. One obtains all of them when one substitutes for t in an arbitrary function of the a_i (law of forms), further all of the consequences from the equation $f(a_1 \dots a_n) = 0$ that concern the a_i, these are the formulas (laws of consequence), and finally the equations from which the given $f(a_1 \dots a_n)$ follows. Löwenheim gave an overview of all the formulas for the solution of the general equation in n unknowns and determined their number, and solved even some other questions of this group.

One arrives at problems of another nature by a transfer of the problems of the algebra of numbers to the calculus of classes. Thus, one understands by a logical ⟦page ends without continuation⟧

⟦7. Metamathematics⟧

47-529

1. Problems made precise through the development of foundational research, exact language

2. Analysis of inferences and axioms (examples of inferences)

3. Different systems (each still extensible)

4. Belief in definiteness with respect to decision for these systems (number theory, analysis)

5. Result that there exist none definite with respect to decision and that there are always undecidable arithmetic propositions at hand

6. Characterisation of formal systems, metamathematics

7. Formalization of metamathematics within the system

8. Setting-up of the undecidable proposition, and proof

9. What do undecidable problems look like **?**

10. Undecidability of freedom from contradiction, consequences for Hilbert

11. Construction of a hierarchy of systems, but none is definite with respect to decision

---

The development of foundational research has allowed many a question ⟦added above: problem⟧ to be formulated in a precise way and to be treated mathematically, questions that earlier had to be left to subjective opinion and philosophical speculation. One such question is, for example, whether each precisely posed mathematical problem is solvable, or whether there perhaps exist unsolvable mathematical questions.

To be able to enter more closely into these problems, an analysis of the proof methods used in mathematics is required in the first place. This analysis was carried through by Frege, Peano, Russell, and others, and its result was that one can lead back all of the ways of inference applied in the whole of today's mathematics to a few axioms and rules of inference. A precondition

47-530

for this analysis is the creation of an exact formula language in place of the imprecise word-language, as delivered by logistic.

All these attempts at the construction of an exact language and an analysis of the ways of inference with Frege, Peano, Russell, and others, went originally in the direction of creating a unified formula language for the whole of mathematics and for the axioms and rules of inference that belong to it, in the sense that each mathematical proposition at all thinkable is expressible in this formal system, and each proof capable of being carried out by these axioms and rules of inference. Such has, however, not been reached. It is admittedly possible to specify systems that contain the whole of today's mathematics and infinitely more, but it remains always possible to indicate certain concept formations that go beyond these systems.

⟦8. General metamathematics (Princeton)⟧

19-792R

⟦The page begins with five items, 4–5 moved to later, 1–3 cancelled:

1. Antinomies, two kinds, first and second kind.

2. The concept true defined, applied to different systems, separation of ever higher types and theorems about it.

3. Repeated construction of undecidable propositions and proof of the unprovability of freedom from contradiction and decision of the propositions through the next-higher types.⟧

A. Antinomies of the first kind ⟦added above: limitation of naive logic remains a matter of taste⟧ type theory the only? (cf. below)

– Short presentation of Russell's type theory

– Extensions: 1. Mixing of types 2. Each proposition true or false 3. Continued into the transfinite

B. Antinomies of the second kind ⟦added above: dirtiness⟧: No other concepts in principle than the above logical, and empirical.

If one hits upon the concept of a true proposition, a definition can be given, and will be given.

Transition to formal systems, exact definitions, and the property $W(a) \equiv A$, $Bew(a) \rightarrow A$

C. Construction of undecidable propositions and proof that freedom from contradiction unprovable.

Further, proof of freedom from contradiction in higher systems – one must state, to this purpose, what recursive.

4. Decision problem, non-constructive proofs of existence, limit in the length of proofs.

5. The axiom of reducibility, the system of Church.

19-793L

Remarks:

1. Set = property

2. Ramified type theory not dealt with, real numbers no difficulty, advantage removed by the axiom of reducibility.

3. Disadvantage that no mixed sets possible.

4. Unclosedness ⟦Unabgeschlossenheit⟧ reflects the existence of undecidable propositions.

5. The concept of "set" is only apparently contained in the axiomatics of set theory.

19-793R

One can divide the antinomies of set theory into two sharply delimited classes (as first clearly spelt out by Ramsey). The first one comprises the purely logical antinomies, i.e., more precisely, they involve only the following purely logical concepts: 1. Concepts of the propositional calculus ( ). 2. All and *E* exists. 3. The concept "property" (propositional function), or "set" and with that the concept that belongs to it, an x is "contained" in a set (the ε relation), and the property F "belongs" to x. To this group of antinomies belong, for example, Russell's antinomy, that of the greatest cardinal number, and the one of Burali-Forti.

The second group (epistemological antinomies) is distinguished by its involving, next to the logical concepts, the semantical "meaning relation." It is the relation that obtains between two things a, b when a is a symbol and b an object that it denotes (in a specific language).

The resolution of antinomies of the first kind consists in a limitation of what is called "naive logic" so that it does not allow of the formulation of the antinomies anymore. This can clearly happen in different ways and there is no possibility to distinguish among these different ways one as being "correct," against the others "being wrong," because the distinction into correct and false already presupposes a logic. Even if that is so, it can still be said that only one resolution has been "successful"[1] in the sense that it avoids the antinomies, but maintains the rest of classical mathematics and set theory with their full scope, and that is "type theory." It has appeared, historically, in two different forms, one of them presented

[1] by ⟦no reference given⟧

in the *Principia Mathematica,* the other in axiomatic set theory (by Fraenkel).

19-794L

Let it be explained now why the axiomatics of set theory can be considered a natural further development of type theory. As is well known, the limitations that Russell's type theory brings to "naive" logic consist in eliminating from logic the concepts "set," "property," as well as the concept "object" that would have to comprise both sets and properties (understood in the most general sense), and thereby also the variables that run through all properties or sets, and those universal and existential propositions that relate to sets or properties. To be able to talk about "sets" or "properties," a domain of things *A* must instead be given (empirically or constructively), on which one can form the concept "arbitrary set of elements from *A*."

19-794R

Hereby one has obtained a new domain of things, "set of elements from *A*," on which one can apply the same procedure, by which one obtains the concept "arbitrary set of sets of elements from *A*," or "set of type 2 of elements from *A*," etc. It follows further that there can be only variables that run through a specific type, assuming? that all- and existence-signs are allowed to refer only to specific types.

The formal system that Russell obtains[2] shows, however, many more restrictions than would be required by this idea for avoiding paradoxes, namely:

1. Russell avoids the formation of what are called mixed types. These are sets that contain elements from two different types, e.g., sets that contain as elements individuals as well as sets of type 1.

2. Russell iterates the process of formation of sets of ever higher types, as described above, only *finitely* often, i.e., there is a specific natural number *n* that belongs to each set that is at hand in his system (as its type). Nothing prevents, however, to sum up under one

[2] One finds a short presentation of this system in ⟦presumably Carnap (1929)⟧.

concept (sets "of finite type") each set that is obtained in his system, and to start the whole process from the beginning, by which the sets of finite type play the role of individuals – therefore give rise to the concept of "arbitrary set of sets of finite type," or set "of type ω." A set of type ω will then contain, in general, as elements sets of type 1 as well as 2 as well as 3 (and further for each natural number n). It can be seen from this that the putting apart of the second restriction carries with it the same for the first.

19-795L

3.) The proposition $x \varepsilon y$ is sometimes neither true nor false in the *Principia Mathematica*, but senseless, namely in case x and y are not of the appropriate types (for example, when x is of a higher or equal type with y). Even this restriction can be easily removed, through the definition of a new ε-relation such that $x \varepsilon y$ is either true or false for arbitrary pairs x, y, namely true if $x \varepsilon y$ holds for the earlier ε-relation and false if $x \varepsilon y$ for the earlier relation is either false or senseless.

The system of *Principia* is not essentially extended if restrictions 1 and 3 are left out, i.e., one can show that one can (as can be easily proved) associate to each proposition s of the new system a t of the old one so that if s ⟦t⟧ is provable in the old system, then t ⟦s⟧ in the new one and the other way around. It follows especially that no contradiction can arise through the leaving out of 1 and 3. It could seem at first that the Russellian paradox could appear again in connection with 3. This isn't so, however, for let x be a variable that runs through all sets of a finite type. One can build without further ado the set of those x for which $\sim (x \varepsilon x)$, and this set comprises all sets of a finite type, i.e., it is the universal set of type ω. It is called Λ_ω, and we have, then, $x \varepsilon \Lambda_\omega \equiv \sim (x \varepsilon x)$ for all sets x of finite type. Λ_ω though, is no such set and cannot, then, be substituted for x. The leaving out of 3 instead, delivers a real extension of the system of *Principia Mathematica*, and there exist especially (as will be shown below) propositions of the old system (even arithmetic propositions in the sense of), undecidable in the old system, that become decidable through the addition of the type ω. There remain, however, arithmetic propositions that are not decidable in the type

ω, but only in

19-795R

the type $\omega + 1$, and so on.

B. We have constructed in the first chapter a logic, or rather an infinite sequence, that avoids antinomies of the first kind and maintains classical mathematics to the full. We have emphasised there that there is no possibility to distinguish this logic as "the correct one" in any objective way, because the distinction between correct and false already presupposes a logic. If, however, we have made up our minds about this special logic, then there is a criterion for when a proof is correct and when it is wrong, and it must be possible to resolve the antinomies of the second kind in the sense that one shows a logical error in them, (namely a violation of the logical rules on which we once decided), and this succeeds in fact.

Part III: Gödel's reading notes

J. von Plato, *Chapters from Gödel's Unfinished Book on Foundational Research in Mathematics*, Vienna Circle Institute Library 6, https://doi.org/10.1007/978-3-030-97134-2_3

1. Editorial remarks

When Gödel began his joint book project with Arend Heyting, he prepared a list of works that is found in an unnamed *Heft*, reel 20, frames 499–526. John Dawson's description of this notebook reads: "Mathematics notebook: Undesignated, Both covers missing, Binding disintegrated, no date." It consists mostly of just names and titles, often lightly cancelled perhaps to indicate that he had seen an item. Some do have summaries, from a few words to more than a page, notably the works of Ramsey that are absent from the systematic *Altes Excerptenheft*. Often an item is listed with a short title or description, then other works of the same author get listed in the style of "Ann 101" or "Tr 17." These can be figured out, but it is not possible always to figure out what use Gödel has made of them. One finds, even if very rarely, Gödel's opinions on his sources, such as "nichts wert" or "unintressant" (nothing worth, uninteresting). The first page of the disintegrated notebook is about the work of C. Langford, meticulously refereed, but nothing critical is said, when instead Martin Davis described Langford as "badly confused" in these very same papers[1] – a judgment that must have been clear to Gödel as well.

The untitled notebook has the appearance of a comprehensive list of possible sources. As mentioned, the sentences are usually incomplete. What turn out to be direct or close to direct quotations can have crucial words missing. Whenever I have found the corresponding place in the paper in question, I have completed accordingly, but a thorough reading of some 130 papers and books would have had the same consequence it had for Gödel, a work left unfinished. My reading, therefore, has an amount of guesswork in it, but what Gödel is writing about is as a rule clear.

The *Altes Excerptenheft* is more carefully written. It consists of references and summaries, often also formal developments, that are direct preliminary work for the book on foundations. I shall give most its contents to the point, at page 95, where the part written many years later begins. Many of the summaries are of a discursive

---

[1] American logic in the 1920s. *The Bulletin of Symbolic Logic*, vol. 1 (1995), pp. 273–278.

nature and contain Gödel's views, but often there can be a plain summary of contents, or just a name and an indication of a journal or title. There are formal developments at places not reproduced in the chapters Gödel wrote.

A comparison of Gödel's list with the extensive bibliography found in Fraenkel's 1928 book on set theory shows that the latter was one of Gödel's main sources for his initial list.

Next to the two notebooks, there are Gödel's lists of names in the second notebook and three pages of "names and dates" in them. The former consist of names and pages and their contents are covered by the lists of names in this book. The latter three pages contain some groupings of various themes such as "propositional calculus."

There are further six separate pages, frames 533–536, with the title "Literatur über Intuitionismus" found on frame 533. These pages have been folded, and many more of the same kind are found in reel 32, frames 76–120. These loose notes would seem to be Gödel's original lists from library visits, with even items on formalism and intuitionism, the topics covered by Heyting, included. Gödel was working also with intuitionistic logic at the time, for his contributions to Hahn's seminar, which should explain these items.

I have collected all of Gödel's sources into a single bibliography and give in the text of the unnamed notebook a name and a year. What these sources are is clear for the said notebook, but for the *Altes Excerptenheft*, it remains often open which particular work has been used. The comprehensive bibliography of Church, comprising all of the fourth issue of the very first volume of *The Journal of Symbolic Logic*, is the best device for narrowing down the alternatives in cases where Gödel mentions just a name. That bibliography and the references Gödel gives in his reading notes also indicate what the papers are Gödel is thinking about in his own book where typically just names are given, with space left to be filled in by a reference.

Gödel's meticulously preserved request slips for libraries provide dates for when he worked on what topics. The dates can be found in the bibliography. A letter^{D} after a reference in the main text tells that there is a date at the bibliography. These dates show, incidentally, that Gödel worked very hard on his part of the book project in the second half of the year 1932.

2. The untitled notebook

1

Langford (1927a) (second paper)

Decidable axiom systems: 1. Linear dense order, a. with a first and last, b. without a first or last, c. with a first but no last ⟦element⟧.

It is proved that the truth value of each proposition of the first order is determined (though not formally decidable, but just through contentful inferences). (Identity is used in a contentful way, so basic concepts $R, K, I,\ K$ the class that is ordered.)

Proof of completeness in the same sense for a discrete order with a first and no last element, where each element has an immediate successor.

Langford (1926b). Axioms of order for a relation R together with axioms of betweenness are complete relative to general laws, i.e., (under the assumption that the domain contains a definite number n of individuals), each general law is either derivable from the axioms or contradicts them. It is shown that each number $1, 2, 3, \ldots n$ of elements has a *specific form*, i.e., there exists a permutation under which it satisfies certain relations.

Langford (1927b). The definition of inductive numbers is rejected, because it could accidentally be the case that there is no property that does not belong to a_ω and that is finite; beyond that, impossible because of the type.[2]

Distinction into modal and non-modal variables. Modal variables need not have any specific order. Claimed that here, a proposition

---

[2] ⟦Langford's point is (p. 605):"It might be supposed that we could express the fact that a_n is an inductive distance from a_0 by saying that a_n has every hereditary property which belongs to a_0; but this is not the case. For example, in a series of the form $a_0, \ldots, a_n, \ldots, a_\omega$, where a_ω is the limit of the progression, it might be true that a_ω has every hereditary property which belongs to a_0; whereas it is of course false that a_ω can be reached step by step from a_0, and false that $f(a_0)$ and f is hereditary entails $f(a_\omega)$."⟧

can be correct on logical grounds. Inductive numbers defined as in Carnap (all of it in connection with Taggart).[3]

Modal variables don't contain the totality of values, therefore no determination of ordering.

2

Hardy (1929)^D. Example of an incomplete symbol ∞ in $\int_0^\infty$.

Church (1925), p. 318

Church (19??) Annals p. 68

Zylinsky (1925)^D, On the theory of deduction.

3

Tajtelbaum (1923), Hahn's question.

Tajtelbaum-Tarski (1924)

Tarski (1924), definition of finite sets.

Nicod (1916)^D

4

Ramsey (1929). A formula of the shape $(x_1 \ldots x_n)F(= \varphi\chi \ldots x_1 \ldots x_n)$ is either satisfiable in each domain of individuals, or there exists a natural number n such that satisfiable for each $k \leqq n$, $> n$ not satisfiable.

One can determine, for each formula, a number h such that for each $n \geqq h$, satisfiability in a domain of n individuals is equivalent to the (decidable) condition B. By $n \leqq h$ only finitely many trials needed, so the above number can be determined. h grows very fast. If there is only one two-place function variable in F, $h = 2n!!!$ [with three-place unknown].

Case of a property satisfiable if and only if satisfiable (in each domain $n \geqq h$) through: the universal function, the null function, identity, difference, serial relation ($\stackrel{+}{-} =$) where only one serial relation to try!

Extended to the case that an E-sign stands before an all-sign.

---

[3] ⟦A reference to J. McTaggart's note *Propositions applicable to themselves*, in *Mind*, vol. 32 (1923)^D, pp. 462–464.⟧

Hertz (1922), (1923), (1929)^D. On axiom systems of the propositional calculus.

5

Wernick (1929) Independence of the second distributive law, decision procedure for a partial system.

Bernstein (1929) Axiom system for the propositional calculus.

Bernstein (1931) Russell.

Bernstein (1916) Here also a general procedure about realizability of system of Boolean algebra.

Bernstein (1932) Whitehead's criterion for the uniqueness of solutions of logical equations proved.

Curry (1929)

6

Dubislaw (1931)

Bolzano (1920)

7

Lindenbaum and Tarski (1926)^D, (1930)

Grelling and Nelson (1908)^D

8

Pasch (1919), (1914), (1919)^D, (1927)^D

Richard (1907) On Zermelo's axiom, (1903) (book), (1920), (1906)^D

9

Study (1929)

König (1905) Distinction into finitely definable and not. The finitely definable ones are denumerable but this enumeration is not finitely definable, and similarly in Borel's *Leçons sur la théorie des fonctions*, Note IV. Therefore ideal elements in the continuum are necessary, for which only freedom from contradiction provable.

Dixon (1907)

Hobson (1907) The distinction ⟦in König⟧ has no absolute sense but at most a relative one of what can be defined starting from certain concepts.

Russell, type theory, *Principles* pp. 270–281 real numbers, (1910), (1908)

Antinomies *Russell* (1906c)

$(RS)\ R\ T\ S \equiv \sim RRS$, the smallest undefinable ordinal number (is defined by these words). The totality of *all* propositions, names, definitions, classes, ordinal numbers is incorrect.

Whatever necessarily contains all of a totality, i.e., what cannot be defined otherwise, cannot be a member of this totality. Totality is something that allows to speak of all of its members.

10

Erkenntnis, vol. 1: Fraenkel (1930)

Stamm (1922) On logic and arithmetic.

Kobrzynski (1930)

11

Hölder (1924)

Couturat

Poincaré (1912)^D On freedom from contradiction. (1909), (1906a)^D (sections VII and VIII)

12

Lewis System ?? strict implication, Hahn.

Sheffer (1913)^D

Löwenheim (1910), (1913), (1915), (1919)

13

Moore (1910)^D

Russell, Principles p. 270 real numbers, p. 323 antinomies, p. 305 proof that each set has a cardinal number.

Post (1921b)

Whitehead (1903) Logic of relations, logical substitution groups, cardinal numbers, (1904) Definition of multiplication for infinite cardinals, (1901) Overview of symbolic logic. ⟦A few formulas follow.⟧

14

Cantor (1883) Foundations of set theory. Number of a set = the general concept that each set has in common with all sets similar to it.

Veblen (1906), (1904)

Chwistek (1929)^D, (1932)

Weiss (1932) Criticism of *Principia.*

Miller (1932) The modality necessary, rules of Aristotelian logic.

Fraenkel (1930) Cantor ca. 1880 foundation, acceptance gained by 1900, at this time, the antinomies surface. Exact foundation of analysis, Cauchy, Bolzano, Weierstrass.

15

Discussion between Russell and Poincaré, *Revue de métaphysique et de morale* 1904–1906^D.

Lietzmann (1923)

16

Huntington and Kline (1917)

Huntington (1924)

Huntington and Rosinger (1932), similarly the concept of separation on a curved closed line.

Schönfinkel, Bernays, Curry. Two papers, *Bernays and Schönfinkel* (1928)^D, *Curry* (1930).

17

Hobson E. W. claims there are more than denumerably many definable real numbers.

Principles, § 103, 104 zig zag theory, § 344 paradox of the set of all sets.

Burali-Forti (1897), paradox found 1897.

Frege (1893)[D]

von Neumann (1931), *Lezniewski* (1931), *Lindenbaum* (1931) pointedly about exactness of von Neumann's modification procedures.

Kaufmann (1931) Rejection of the higher function calculus, distinction between sets extensional and intensional (numerical specification?), rejection of the non-denumerable. Completeness and decidability follow from the definition of the number series, therefore trivially recognized?

Peano (1906) About antinomies, indication of a distinction into two kinds.

Peano (1897) Axioms of logic.

18

Ramsey (1926) Mathematical theorems are not to be defined as such true propositions in which there occur only logical constants and variables (counterexample: each two things differ in three ways), but it is further required that they are tautologies.

There *follows a definition of tautology* (that Russell did not accomplish and that comes from Wittgenstein). Wittgenstein recognized that one can use this definition also on infinite propositions.

> Each proposition expresses accordance with some combinations of truth values of the elementary propositions and failure of accordance with the others. In the case of infinitely many, propositional functions serve to gather together this infinite set.
>
> Each proposition is a truth function of elementary propositions (Wittgenstein).

Reducibility axiom is a matter of brute fact.

Structure of mathematics:

1. Difficulty: Mathematics deals with extensions (the question is not of classes and relations being defined through predicates). This calculation with extensions must be reduced back to a calculation with truth functions.

2. Antinomies distinguished into two kinds (epistemological or linguistic).

Errors in the *Principia:*

1. Class = definable class. Whereas the question whether all classes are definable is an empirical one. This is the ground for why the multiplicative axiom is not a tautology in the *Principia.*

19

2. Ramified type theory and the axiom of reducibility.

With these, mathematical induction and Dedekind cut impossible.

3. Identity and agreement on all properties are only empirically equivalent, therefore Russell's definition not usable. Identity can be replaced by symbolic conventions. This leads to difficulties, because of the axiom of infinity. $x = y$ is, anyway, not an elementary propositional function.

Axiom of reducibility, difficulties: Lack of symmetry between a function over individuals and a function over functions. The domain of objects that are substituted for individuals is determined through their sense (they are names for individuals). That is objective (dependent on our method of symbolization, maybe we don't have names for some individuals). Functions instead just ones that can be described in a certain way (this the way of the *Principia,* the subjective method). Objective method here, and also those included that perhaps cannot be described at all. A predicative function of an arbitrary order is a truth function of atomic functions of this order that have the form $\varphi(\ldots\ldots)$. Here the arguments are filled with functions of a lower order that are already defined, with φ the variable for the argument of highest type of this function. Axiom of reducibility holds then, for example, $(\Phi)f(\Phi(z), x)$. That $(\varphi)\varphi a = F(a)$ in which F is included in the domain φ, is analogous to $p\,.\,q \equiv p\,.\,q\,.\,pq$.

The distinction into elementary and non-elementary in the *Principia* is a distinction in expression (the same function can be expressed as elementary and as non-elementary, definition of *the same* led back to *the same proposition*).

20

Johnson (1922)^D Vol. 2, p. 159, something about extension.

Kuroda (1931)^D, p. 33

McTaggart (1923)^D (Propositions that speak about themselves).

Dubislaw (1931) *Die Definition.* 1. Definition = prescription of the denotation of a sign (rule of substitution).

Requirements: 1. Must be eliminable. 2. Must not destroy the formal laws. A formal criterion is put up which states that a proof about? truth, of freedom from contradiction, cannot be destroyed through the introduction of new signs.

Behmann (1931)

Aristotelian inferences (therefore also *modus barbara*) limited through a prescription ⟦Behmann, p. 42⟧:

> A variable cannot without further ado be considered as running through all of the domain that comes, by the symbolic connection immediately at hand, into question, but rather as running through it only insofar as the results of substitution can be written without abbreviating signs.

[The Aristotelian inferences use a process that can make the translation back impossible.]

Even epistemological antinomies are to be excluded (ultrafinite logic).

21

Hölder (1931) The freedom from contradiction of arithmetic can be founded only on a consequent construction of its concepts (otherwise circular). Geometry = science of space.

Curry (1931) (About the universal quantifier, an attempt at the substitution principle) from? formally simple things.

Foster (1931)

Ramsey (1926) and other papers, continuation of the treatise (1926) ⟦1926-7⟧.

Resolution of the heterological antinomy:

$$F(x) \equiv (\exists\varphi)xR(\Phi\hat{x}), \quad \text{“}F\text{”}RF(\hat{x})$$

does not hold, because R (the meaning relation) has different levels that cannot be summarized into one. Thereby there follows a division into levels of "linguistic expressions" which, however, do not belong to the propositional functions themselves. *Propositional functions in extension* are introduced because of the undefinable classes, axiom of choice, identity (this is needed already for having the finite ones, but it is no elementary propositional function).

By maintaining predicative functions, we have 1.) a misinterepretation (especially the identity) 2.) the axioms of choice and infinity empirical questions.

22

Tarski (1930)^D $F(x)$ = those consequences of $x + B$ that lie in A. Then the axioms are satisfied, there is for each system a sequence of propositions x_r that are "independent after level" and a basis for the system, possibly infinite.

If a monotonely growing sequence is axiomatizable, there exists a sequence system that is equivalent to the first sequence.

An infinite set of "propositions independent after level" is not axiomatizable.

S axiomatizable ≡ there exists no infinite sequence independent after level that is equivalent to S

≡ there exists a finite sequence independent after level that is equivalent to S

≡ there exists a monotone strictly growing sequence of systems that is equivalent to S

If there exist infinitely many independent propositions, the set of axiomatizable systems = $\aleph_0$

The non-axiomatizable ones = $2^{\aleph_0}$

A set of propositions is called *complete* if each meaningful proposition is a consequence or its addition contradictory.

Each set of propositions *free from contradiction* can be supplemented into a compete system (trivial).

23

Cardinal level of completeness = number of systems $g(A)$ that contain A.

Ordinal level of completeness = smallest ordinal number for which no ascending sequence of systems free from contradiction that contain A, $\gamma(A)$.

$$A \text{ free from contradiction} \equiv \text{Ord} > 1$$
$$\equiv \text{Kard} > 1$$
$$\text{complete} = 1 \text{ or } 2$$

⟦There follow some inequalities for $g(A)$ and $\gamma(A)$.⟧

In part II proved ⟦as announced at the end of the paper, no part II exists⟧:

γ either finite or Ω

g either finite or $2^{\aleph_0}$

Ramsey ⟦(1926)⟧ continued:

Function in extension is explained as an association of propositions to individuals (it is not required that $\varphi\, a$ about a the same proposition as $\varphi\, b$ about b). Justification is an intelligible notation giving a definite meaning to the symbols.

24

New interpretation of the *Principia* (form is maintained), tautological inferences. By predicative functions of individuals are herein to be understood functions in extension, and by predicative functions of higher types the predicative functions in the earlier sense. (The counterexample by Bernays for the axiom of reducibility, an axiom system for $>$ in the domain of natural numbers, i.e., all the relations that can be achieved from it are not sufficient for the definition of inductive numbers.)

Axiom of choice in the sense of the *Principia* no tautology, for there must exist predicative choice functions and axiom of infinity not satisfied, axiom of infinity because $x = y$ always either tautology or contradiction.

Even propositions about cardinal numbers either tautologies or contradictions, because $<$ different in different worlds, so then not provable from the general logical axioms, but instead as a new axiom. Axiom of choice, even if tautology, likewise as a new axiom because proof would be infinite.

Russell, Vicious circle principle, *Principia,* first edition, introduction, ch. II, p. 37

What presupposes all elements of a set cannot be an element of that set.

If from the presupposition that something builds a totality follows that it contains elements that can be defined only in terms of this totality, it cannot form a totality. To form a totality = one can speak of all.

φ cannot belong to the arguments of φ, because it presupposes all elements for it to be determined.

⟦Addition on next page: Another justification: A function is an ambiguity and must therefore occur in propositions so that the ambiguity gets the determination needed (e.g., $(x)\varphi x$), therefore never occurs like an individual.

Definition. Function = something that relates in an ambiguous way to the elements of a totality, p. 39.⟧

There exist different kinds of truths. Truth of first order for elementary propositions = propositions without apparent variables (therefore without variables),

25

i.e., there exists a corresponding complex. Truth of second order, form $(x)\varphi x$. Even negation and disjunction have a different sense for non-elementary functions (10*).

Grounds for assuming the axiom of reducibility: The grounds for each axiom are inductive, i.e., propositions follow from it that are probably true, no other true ones known from which these propositions follow, and no propositions follow that are probably false. Self-evidence is just one of the grounds for the assumption (that is equally justified with the others), i.e., he treats the axiom of reducibility like a physical hypothesis – this the general view at the time, cf. preface, p. V.

Russell, Principia, second edition after Bernstein's ⟦(1931)⟧ summary.

Modifications. 1. Apparent variables and the assertion sign before a propositional function left out.

Axiom 1·11 left out.

Nicod, Wittgenstein (reducibility).

Already in the first edition: The known methods for proofs of independence are not applicable on the fundamental theory.

Russell the first one to define real numbers *"as cuts."*

Intensional propositions: In A believes p, p is a factual proposition.

In $p \supset p$, p is an assertoric proposition.

26

The first p is not identical to the second. In the first case, p is used to talk about something, in the second case, p is spoken about. In the first case, p has a "transparent quality."

In the first case, p is a singular occurrence, in the second a class of occurrences.

In the first case, it is not p but just the components of p that occur, therefore one cannot infer:

$p \equiv q . \supset . p = q$, because p, q occur in the second part in a sense different from the first.

Couturat (1914), *Venn* (1881)^D, symbolic logic = calculus of classes.

Formulaire ⟦Peano (1897)^D⟧

⟦Summary of Peano's notation for classes, membership, operations.⟧

27

Arithmetic Dedekind ⟦(1888)⟧ If, in considering a simply infinite system N ordered through a mapping φ, one disregards completely the specific constitution of the elements and just maintains distinguishability, and considers only the relation to which they have been set against each other by the order-mapping, then these elements are called natural numbers.

Formulaire ⟦Peano (1897)⟧

$+a$ is considered, without being separately defined, as an operation, because $x + a$ is defined.

Negative and rational numbers introduced as operations, $/a$.

Distinction between operation and sign for an operation.

Character of the definitions (notations ?)

I $x = a$ e.g., $Np = (N+1) - [(N+1)(N+1)]$

II $h \supset x = a$

x would be a combination of signs that possibly contains also known signs, or only known signs, and sense is given only to a certain combination of these signs. Here, a rule is also given by which the *definiens* must contain as many variables as the *definiendum*.

$a, b \,\varepsilon\, N \supset quot(ab) = max(No \cap \overline{x3}(xb \leqq \omega)$

III Through abstraction, $\varphi(a) = \varphi(v) = p_{av}$ Df

φa is what one obtains when one considers in a only the properties that it has in common with all v for which $\varphi(a) = \varphi(v)$ holds,

28

for example, relations in Euclid, rational numbers – objects one obtains from pairs of numbers when one considers only those properties that are invariant with respect to equations (Stolz), real numbers.

$a, b \,\varepsilon\, KR \supset l'a = l'b = \ldots$

Length of two line segments (when through movement to coincidence), parallelism of two lines (calculation of points infinitely distant).

Vector *cardinal number* *ordinal number*

Notations de logique § 41: Definitions are abbreviations and therefore dispensable.

Choice of primitive ideas in whatever science is arbitrary. One determines, or if you want, one defines the primitive ideas with the help of their fundamental properties. *Transformation of proofs into logical signs = analysis and resolution into single steps (this is much more difficult than the resolution of the propositions).*

Tarski Tajtelbaum (1924a, 1924b)

Russell, Principles, pp. 16–18, $\supset$ as the sole basic concept.

$$\sim p \equiv . p \equiv (q)q \qquad Df \sim$$

$$p \supset q . \equiv : p \equiv . pq \quad Df \supset$$

$$p \vee q \equiv . \sim p \supset q \qquad Df \vee$$

$$pq . \equiv :. (f) :. p \equiv : (r)p \equiv f(r) . \equiv . (r)q \equiv f(r)$$

29

⟦There follow formulas to page **31** that deal with the role of the extensionality thesis, in reference to *Principia*, section I.A.∗2.⟧

31

von Neumann (1928)[D]

Dedekind (1888), § 126, § 130

In set theory, the finite numbers can be built up without the condition that there exists an infinite set.

$y = \{x\}$. Proof that there exist infinitely many sets, but one cannot form the set of natural numbers. Also complete induction introduced in this way **?** Partial realization in von Neumann (1928b)[D].

Peano continued:

Le problème proposé par Leibnitz est donc résolu. We are in a position to express each mathematical proposition with the help of a few signs that have a precise meaning and that obey precise rules. (Aim of the *Formulaire* is the ordering of the bibliography of the different areas.)

Difference between individual and class:

$$\left.\begin{matrix} x \, \varepsilon \, a \\ a \, \varepsilon \, \iota a \end{matrix}\right\} \text{ different sense} \qquad \left.\begin{matrix} a \cup b \\ \iota a \cup \iota b \end{matrix}\right\} \text{ different sense}$$

32

In place of an abstract theory of relations, he has one for functions.

The propositional calculus:

⟦Axioms and rules for the propositional calculus follow.⟧

Proofs are not carried through in a precise way, but can be constructed.

The calculus of classes:

Primitive ideas K, ε

Pair = (x, y), brackets, variables

Difference between apparent and real variables clarified, first primitive idea $\supset_{xy\ldots}$, the indices left out when $\supset$ is the sign for implication. Peano also introduced points, together with a$\supset, \sim$ as basic concepts.

33

⟦There follows a list of Peano's basic concepts.⟧

(The first symbolic treatise of Peano: *Arithmetices principia, nova methodo exposita*, 1889[D].)

⟦Some of Peano's axioms follow.⟧

34

Peano, general characterization:

Aims: 1. A precise language. 2. This a means for discovering the basic concepts of axiomatic theories. 3. Brevity, presentation of theorems in a smallest possible space, and perhaps also ways to discover new ones. The first contains in fact 3 and 1 (2 neglected, the points come from Leibniz and were used first by him).

Distinction into ε and $\subseteq$, analogous to a and ιa. It is a merit that the relativity of the basic concepts and axioms has been emphasized.

35

The Weierstrassian theory of irrational numbers presented in Mittag-Leffler (1920).

A number is a group, i.e., a denumerable set of rational numbers such that the sums of finite subsets are limited.

Cantor (1874), Heine (1872), Méray (1872) at about the same on the concept of a fundamental series, i.e., $\lim a_i,\ a_{i+1} > a_i$.

Dedekind (1872), (1888) gives in the preface ⟦of the former⟧ a purely arithmetic definition of irrational numbers.

36

Lipschitz (1877)

Méray (1872). Irrational numbers are fictions that serve to express theorems about convergence in an invariant way.

Heine-Cantor (1872, 1874). Their numbers are *signs* associated to the fundamental sequence.

Weierstrass already earlier than 1872, ca. 1860 in lectures that, however, were not published. ⟦This item continues from top of page 38.⟧

Orloff (1928)

⟦A propositional axiom system is given, followed by:⟧

$\rightarrow$ should be interpreted as: follows logically. Further, a theory about *Bew* that rests on three axioms:

$$\Phi a \rightarrow a,\ \Phi a \rightarrow \Phi\Phi a,\ \Phi(a \rightarrow b) \rightarrow \Phi a \rightarrow \Phi b$$

Ramsey (1926-7) Presentation of the actual situation (the three directions and summary of his main work *Foundations*).

Russell defines mathematics as the totality of true propositions that don't contain other than logical constants.

37

That is incorrect, because there could be propositions that are true by accident. The aspect of necessity must be added. This achieved through Wittgenstein's definition of a tautology.

The "orders" of propositions and propositional functions are not properties of what is common but of the symbols one uses to designate something. Propositions of the second order "designate" in a way different from propositions of the first order.

The axiom of infinity: If it were false, then, for example, Dedekind's proof for class numbers would not be applicable, and therefore the proof has to be rejected altogether. Even a proof of freedom from contradiction would be of no help, because the Hilbertian proof of freedom from contradiction holds only for numbers that in fact can be symbolized in the system.

Schweitzer 1. Advantage of superfluous basic concepts is the avoidance of existential postulates.

(2) Kempe's[D] law of continuity is an anticipation of Hilbert's completeness axiom.

(3) [generalization of complete independence]

Stamm (1922) (1) A completely trivial reading of logic through arithmetic operations and some relations.

38

Cantor (1872), (1883) He distinguishes the sizes of numbers ⟦Zahlengrössen⟧ (value or limit) of the λ-th kind, by the iteration of the process of building limits, but even admits that it is only a question of things being given in a different way.

Heine associates the rational numbers as signs to a series that consists of nothing but the same elements.

Kossak (1872), the theory of Weierstrass presented here.

Cauchy, his convergence criterion not yet proved.

Bolzano tries to prove it in "Rein analytischer Beweis etc" ⟦1817⟧ by showing that it can be calculated to arbitrary precision.

Preliminary works Gauss, Lagrange, Cauchy.

Peano (1906) The contradiction of Richard arises from this, that a definition contains its concept "lingua communis," something that is quite familiar but not determined, and it is the basis of the ambiguity, at p. 152, example the smallest common multiple.

Refutation of Poincaré (1906b)^{D}: Russell (1906b)^{D}, p. 632, Zermelo (1908), p. 116 ⟦Addition from page 43:⟧

The same kind of inference in the formation of a minimum. Proof of the fundamental theorem of algebra by Cauchy. A definition *does not create an object* but just characterizes it, and the concept to be defined is just equivalent to one already at hand (not identical). A concept of equivalence occurs, though, always in a definition. (That is what Peano ⟦(1906)⟧ says in *Rivista*, vol. 8.)

Further, some theorems are given:

1. The sums of equivalent sets are equivalent.

2. The sum of denumerably many denumerable sets is denumerable.

⟦End of addition.⟧

39

Russell (1906a ⟦or 1908?⟧), (1905)

An asserted propositional function (with free variables) is no proposition, but a scheme that produces propositions. One can assert propositional functions also for things the totality of which is contradictory ($p \vee \sim p$), but one cannot define, e.g., the finite numbers in the known way.

One can form a totality ⟦added: that uses the word all⟧ if it is the intended domain ⟦Sinnbereich⟧ (or part of such) of a propositional function.

In the usual logic, propositional functions that are in truth divided into different intended domains have a name (e.g., *or*, ε, etc). All ranges always over all those for which it has a sense.

Vicious circle principle: "*No totality can contain members that are defined in terms of itself.*"[4]

Application: "What contains an apparent variable, cannot belong to the domain of this apparent variable."

Type theory: A predicative function is one the order of which is greater by one than ⟦added: the apparent variables have the same or lower type than the arguments⟧ its arguments. (also propositions of the n-th type!) Resolution of the paradoxes.

Impossible to prove: If n, m are finite numbers, then also $m + n$.

The axiom of reducibility is harmless for paradoxes, because it requires only equality of scope.

40

Russell, Principia Mathematica

Basic concepts:

1. Propositional function 2. Predicative propositional function 3. Negation 4. Disjunction 5. (x) 6. Universal propositions through apparent variables 7. $\vdash$ (assert) (= as a definitional sign)

Df. $p \rightarrow q,\ pq,\ p \equiv q, \quad (\exists x), =$

$\sim(\sim p \vee \sim q), \quad p \supset q \,.\, q \supset p$

Axioms:

I Five propositional axioms II $(x)\varphi x \supset \varphi y,\ (x)p \vee \varphi x \supset p \vee (x)\varphi x$

III Rule of implication IV One infers from $\varphi(x)$ to $(x)\varphi x$

[V $(x)\varphi x \supset \varphi a$ for constants a, this principle not needed in pure mathematics, only in applied]

⟦Cancelled: Two primitive propositions about type theory, $\varphi(x)$ and $\psi(y) \supset \varphi x \psi y$ and x, y of the same type⟧

41

⟦possibly left unintentionally uncancelled: that are example propositions for stating that two different types are disjoint⟧

[4]⟦Last four words in English in the original.⟧

VI $(\exists f)(x)\varphi xy \equiv f!xy$

All axioms hold all the time for all types, and the question in each proposition is only about relative types.

Difference between intensional and extensional:

1. Propositions: *A* believes *p*, *A* says *p*, *a* occurs in *p*, *p* is a tautology, *q* follows logically from *p*.

2. Propositional functions: *A* believes that $(x)\varphi x$, $\varphi \equiv \psi$, φ is predicative (i.e., no apparent variables occur in φ), φ and ψ mean the same (i.e., logically equivalent).

Introduction of classes. It is practical to treat $f(\hat{z}\varphi(z))$?? as if it had a single argument $\hat{z}\varphi(z)$. (The argument, is, in truth, $\varphi(z)$, $\hat{z}\varphi(z)$ is a "fictitious object.")

⟦some formulas follow⟧

42

⟦formulas continued⟧

Type theory has, for ordinal and cardinal numbers, as a consequence that one arrives, in any case, at each finite number, up to $\aleph_n$ and ω_n (for each natural n).

(It is impossible to complement the series of ordinal numbers in such a way that they pass over each finite type. The series as a whole is a "fiction.")

Fundamental principle of type theory: Each expression that contains apparent variables is of a type higher than these variables.

Russell (1906c) about the axiom of choice.

43

Poincaré (1909) Objection *Acta Mathematica*, admits that the definition of minimum is not predicative, but the proof can be carried through by forming the lower bound of the rational numbers.

⟦Addition to page 38.⟧

Hahn (1909)^D

Langford (1930)

44 ⟦blank⟧

45

Whitehead (1901), (1903)

⟦A detailed formal summary of Whitehead's calculations.⟧

46–47

⟦Formal details from Whitehead (1903), "The logic of relations, logical substitution groups, and cardinal numbers," continue.⟧

48

Russell (1906c)

The problem consists in finding a condition for when a propositional function determines a class, but even the propositional function itself cannot always be considered as an object (that leads to contradiction). *I.e., a proposition cannot always be separated* in x and φ. One can prove that there exists at least one definite propositional function to which there exists no class.

$(R \,\varepsilon\, K = \sim (ES)RPS)$

All contradictions stem from assuming that there exist classes for propositional functions. Here "predicate" is defined. They are names that determine classes.

General form of contradictions: Φ f (defined), if each member of a belongs to Φ, then $f(a)$ belongs to Φ but not to ω. Russell's antinomy, for example:

$\Phi(x) = x \sim\varepsilon\, x \;\; f(\omega) = \omega$. Way out either

1. Φ determines no class, or
2. $f\Phi$ does not exist.

Each such f produces (under certain conditions for these f) a series that is homeomorphic to a series of the ordinal numbers *the members of which cannot build any class.* f is a selfreproductive process (Φ selfreproductive property). One can give, thanks to the process f, for each set m of things with the property Φ a thing with the property

Φ that does not belong to m.

Two kinds of attempts at a solution:

1. Not each propositional function determines a (separable) class, but only the predicative ones. No class is determined by a.) the circular ones, zig zag, b.) those that comprise the ordinals.

2. No propositional function is an independent entity but just expresses an abbreviation, no-class.

49

Russell (1906c)

Zig zag excludes only circular classes.

Cf. *Principles*, § 103, 104, 484

The axioms are determined only through the desire to avoid contradictions (have no plausibility and are very complicated), for example, there exists a greatest cardinal number but no greatest ordinal number.

Limit of size:

The existence of selfreproductive processes seems to prove that the class of all classes cannot exist.

Definition: A class exists in case it is equivalent to a *proper* section from the series of ordinal numbers, but (thinks Russell) one cannot indicate a criterion for which ordinal numbers exist, up to what point the series is still legitimate.

No-class theory:

Functions are not considered as things that exist in isolation of an argument. Instead of functions, one speaks of propositions p for which through the substitution of certain components $p\frac{a}{b}$ in a specific way, something specific holds. He hopes to develop on this basis analysis and parts of set theory.

Axiom of choice

If the set of subsets on a finite set is finite, then each finite number is inductive. He believes in a *true solution* of the paradoxes and finds it meaningful to ask whether the axiom of choice is true or not (crass realism).

50

Antinomies:

1. Russell (class of all classes)
2. Greatest cardinal number
3. $xRy \equiv \sim xxy$
4. Burali-Forti
5. Liar
6. Heterological
7. Richard
8. Smallest undefinable ordinal number
9. Smallest number that cannot be defined in less than a hundred words

---

Axiom of choice:

1. There exists for each class of classes a relation that chooses an element from each class.

2. There exists for each class of *disjoint* classes a choice class.

How is the multiplicative axiom formulated in the *Principia*?

We are in need of the axiom of choice for showing that the number of elements of α ⟦added above: (disjoint)⟧ classes of β-terms ⟦added: where α is infinite, example: that $\aleph_0$ pairs of socks are altogether $\aleph_0$ things⟧ is always $= \alpha \times \beta$ (in which $\alpha \times \beta$ is defined in the manner of Whitehead as the number of pairs). Similarly for the relations of multiplication and addition.

51

Zermelo class = a class that can be well-ordered, for a class for which there exists a unique mapping of the subsets and the elements contained in them.

Russell already expresses the axiom in the form of *Hilbert:* there exists a function that associates to each class an existing element. The sense of the axiom of choice (and of the doubt raised against it) is always whether there exists a *rule* that chooses from each class an element.

52

Russell (1906c)

p. 41 1.) If [in mathematics] a new entity is introduced, it is not created by the activity of the mind but merely discerned.

2.) Hobson ⟦1906)⟧ claims: number = the characteristic that is common to all sets of the same cardinality, the existence of which is recognized by the mind. Whereas Russell says that there are many such and proves it. *Remark.* Perhaps still unique if one makes it clear that a number Z stands, to all equivalent sets M, in the relation: Z is the conceptual element contained in all the M, and only the M. That which is common to all M, $Z = \Delta_{M\,\varepsilon\,Z} M$

3.) p. 38 zig zag ness. This "would-be-classes"[5] is the property that is best illustrated by: x is a member of w if $\varepsilon\, R'x$

and x is not a member of w if $\overline{\varepsilon}\, R'x$

4.) There exists in the zig zag a universal set V, and with u always $-u$.

5.) Objections to limit of size: The principle does not state when the series of ordinal numbers "begins to be illegitimate."

6.) The basic claim of this theory, the existence of "self-reproductive" processes seems to make impossible the concept of the totality of all things. A set cannot "reach" a certain γ.

---

[5] ⟦Original in English⟧

3. Altes Excerptenheft I (1931– ...)

1

Leibnitz

1. *Characteristica universalis* (for all sciences). (In collaboration with the nations.)

2. General method of inference (in the form of calculation). Puts to test? the advancement of sciences.

ad 1. "alphabet of human thought"⟦English original⟧, only the pure ideas have basic signs, the rest expressed through combinations of these basic signs. The basic concepts are something absolute (as are the basic propositions). Numbers are associated to the basic concepts, and from these (say in the form of fractions), numbers are associated to combined concepts. Basic concepts with prime numbers, combined concepts as products. $S \subset P$ if S is divisible by P.

ad 2. All inferences consist in the relating and substituting of signs.

De arte combinatoria is meant to be a systematic production of new truths, in which one combines the basic concepts in all possible ways (*ars inveniendi*). One has to have an overview of all possible kinds of combinations.

Logical calculus:

1. *Specimen calculi universalis* axioms: substitution of letters by others, exchange of letters in a product *ab*, elimination of repetition.

$$a \subset b \cdot c \subset d \,.\supset.\, ac \subset bd, a \subset bc \,.\supset.\, a \subset b \cdot a \subset c$$

2

$$a \subset a, ab \subset a$$

2. *Non inelegans specimen demonstrandi in abstractis.*

3. Appeared without title in

Gerhadt's *Die philosophischen Schriften von Leibniz*, vol. VII.[D]

Contents: Definition of identity, inference to transitivity and symmetry. $+, -$ as concepts, from these containment defined through

$A + B = C$,[6] transitivity of containment derived from this, $[A + B = B + A]$.

Axioms $A + A = A$, $A - A = 0$, possibility to perform addition and subtraction.

Proved, for example: $A \subset C \cdot B \subset C \rightarrow A + B \subset C$

⟦Proved⟧: There exist sets with arbitrarily many elements that are closed with respect to +.

Lambert and Castillon. Give an intensional calculus (*very* incomplete), no results because of intensional point of view, only Hamilton extensional.

De Morgan. Aristotelian Syllogistic presented with overview and symbolically. Introduces first the concepts *inv. rel.* and *contr.* = a negative relation $-$, *combined*, and proves $-(L^{-1}) = (-L)^{-1}$.

⟦There follow one and a half pages of formal development, mostly about the combination and inverse properties of relations, in particular transitive relations and the indicated conclusion:⟧

4

Theory of transitive etc relations uses for the derivation of syllogisms

$$\begin{array}{c} A\,L\,B \\ B\,L\,C \\ \hline A\,L\,C \end{array}$$

A generalization of the syllogism, in which one disregards the specific meaning of a complex and substitutes for it an arbitrary relation (that is transitive, etc). Further, De Morgan considered as first other combinations of subject and predicate than *a e i o*, though he remains still within the Aristotelian setting of the problem. There stand in particular only three concepts in the premisses. The concept of a *Denkbereich* (universe of discourse) comes from him.

Jevons $+ \,.\, - 0 \quad (a + b, ab, a = b, -a, 0)$

Axioms: 1. Equality axioms (transitive)

[6]⟦Put $C = A$, then if $A + B = A$, B contained in A⟧

2. $ab = ba,\ aa = a,\ a\,.{-a} = 0,\ a + 0 = a$

$a + b = b + a,\ a + a = a,\ a(b + c) = ab + ac$

$a + ab = a,\ a = a(b + -b)$

Rule of inference: $a = b \rightarrow a$ can be put all over in place of b.

Contrary to Boole, each step of a calculation has a meaning with him, $a + b$ also if a, b are not disjoint.

Boole. He works without care, just as in an algebra of numbers, divides, subtracts, even $\frac{0}{0}$. Other results are correct (compare to working with $\sqrt{1}$). $a + b$ is used only on disjoint classes.

Problem with Jevons and Boole: to put up premisses in the form of equations, then to resolve or eliminate, here first the problems of the logical calculus.

Boole is superior to Jevons in mathematical elegance, Jevons is superior to Boole in conceptual clarity.

5

Boole uses already the calculus of classes, after the method of *division* ?

Peirce. ⟦Written at right: On an Improvement in Bs Calculus of Logic⟧

1. Unifies the good aspects of Boole and Jevons.

2. Continues the theory of relations of De Morgan ⟦Written at right: Description of a Notation for the Logic of Relatives^{D}⟧

3. Introduction of apparent variables Π, Σ.

4. Going back to Leibniz, logistic as the general science that lies at the basis of the form of mathematics. ⟦Written at right: Upon the Logic of Mathematics⟧

ad 1. Introduces the relation $\supset$ and states the associative law.

He makes a distinction between material and logical implication in the propositional calculus and states the law:

A true proposition is implied by each proposition. He expresses also as the first the additional axiom for the propositional calculus,

$x \neq 0 \rightarrow x = 1.$

ad 2. ⟦One and a half pages of formulas about the calculus of relations, followed by:⟧

The methods of resolution and elimination for this calculus are not given in full, but a theoretical foundation through *relatives as classes of pairs* defined by them.

7

The concept of apparent variables in the Russellian sense is already at hand in a precise way.

Peirce. Is not satisfied with the putting up of the laws for relatives as axioms, but gives them a foundation through a function calculus with apparent variables. The operations $a + b$ etc are defined, e.g., $(a + b)_{ij} = a_{ij} + b_{ij}$ and $a + b = \Sigma_{ij}(a_{ij} + b_{ij})(I; J)$, respectively, and axioms derived from these. The symbolism Π, Σ.

Gives the rules, in particular the rule of passage, brings to normal form and modifies the matrix (Boolean). Removal of Π, Σ the redex of which does not occur anymore in the Boolean. Π, Σ are conceived of as products, though he emphasizes that these are not derivable from the usual propositional calculus, because of unclarities. If one is allowed to consider numbers in classes as identical, Boolean addition goes over to the arithmetic one. Even arithmetic multiplication, defined as sets of pairs.

Grassmann. Eine Formenlehre oder Mathematik. Puts up a Boolean calculus of classes, beginning with a corresponding law for individuals, $e + e = e$, $e_1e_2 = 0$ etc.

Hugh Mac Coll. The calculus of equivalent statements. Symbolic reasoning.

Builds up the propositional calculus, and independent of Boole.

8

Symbolic Logic And its Applications, a kind of system of strong implication.

Ladd Franklin. Algebra of Logic[D] (Johns Hopkins)

Brings all Aristotelian syllogisms into a common form, namely

$ab = 0 \,.\, c\bar{b} = 0 \,.\, ac \neq 0$

She works with the basic concepts $a \vee b \ldots ab \neq 0$

$a \overline{\vee} b \ldots ab = 0$

Schröder. Principle of duality (earlier doubts by Peirce).

Calculus of relatives built as in Peirce, applied to the theory of Dedekind chains.

Poretzky.

9

Frege before Schröder and *Formulaire* but influence *very* small. Logic is not an end in itself for Frege, Peano, Russell, but a means for building up mathematics.

Lewis. Axioms of the calculus of classes. ⟦One page of formal development in which the six propositional axioms given are said to be duals of those in Huntington (1904). This is followed by Poretzky (1898–9) and an exposition of his system of equations, and formal work on algebraic logic to page **13**. That page continues with:⟧

Propositional calculus ⟦Lewis⟧

[$a \vee b$ means in colloquial language an either-connective (either all persons are mortal or parallels intersect each other)]

Two forms of the propositional calculus:

1. Two-valued algebra = a contentful axiomatic theory with the additional axiom $p = (p = 1)$. Then provable, for example, if either p or q true, then $p + q = 1$ is true and the other way around. The implications $p \supset q \vee q \supset p$, $\sim(p \supset q) \supset\colon p \,.\, \sim q$, $p \supset q \vee p \supset \sim q$ are strange consequences.

14

2. Theory of propositional functions. Lewis makes the assumption that a proposition that is proved of an arbitrary finite number of individuals $x_1 \ldots x_n$ can be applied in general, and he derives from this the theory (not correct in general).

3. Derivation of the calculi of classes and relations from the calculus of propositional functions.

⟦Equations with operations on relations follow.⟧

Strict implication. ⟦Two pages of definitions, axioms, rules of inference, and results follow that end with:⟧

All theorems of the *Principia* follow from this system, and it follows that they are necessary (therefore also for each $P \supset Q$, if it is tautological, then $P \prec Q$).

⟦Page **16** gives two axioms of Becker, mentions E. Nelson (1930)^D, and gives one axiom by Parry.⟧

17

Lewis. Takes the concept to be formalized that states "one can go over from *a* to *b* through a valid inference." This is an intensional relation = any situation in which *p* is true and *q* false is impossible.

Each proposition follows from an absurd one is not justified, and a necessary proposition is implied by each proposition.

Kempe. Derives logic and geometry from common postulates for relations. *Proc. London math. Soc.* XXI (1 Series, 1891, p. 147)^D. ⟦The postulates are given, followed by:⟧

Royce. Introduces instead as basic relations, *Trans Am.* VI ⟦The relations are given, followed by a description of Whitehead's universal algebra in formal terms.⟧

19

Zylinsky^D. All possible 16 truth functions in two variables are investigated. It is determined for each how many can be defined from it, and it follows that *just* Incomp. and a unified falsity suffice to express everything. As shown, this is possible through a finite procedure.

Poincaré 1905^D. Criticism of Couturat's *Principes de mathématique.*

Thesis: *Mathematics is not reducible to pure logic, but intuition plays instead a role in it.* This implies in principle complete induction and the concept of natural numbers.

20

I 1. The definitions are circular to the extent that the concepts "one," "two" are already presupposed in logic. A relation has two terms, logical product of two propositions, there exist two truth values, in Russell's definition of "1" there occurs the number 2.

2. The logical axioms are synthetic judgments *a priori*. (They cannot be any hidden definitions, because otherwise a proof of freedom from contradiction would be necessary, something that is not possible because of the necessity of induction.)

1. + 2. Things have just begotten a different name, but then intuitions are necessary as before.

II The Russellian foundation is not sufficient to ground induction, because numbers are defined implicitly by the Peano axioms (in Poincaré's opinion), and therefore a proof of freedom from contradiction is necessary that can be delivered only through induction.

Frege, letter to Peano. About the incorrectness of his definitions, *Rivista* VI, p. 53.

Russell. Principles of Mathematics, § 17. The known methods for proofs of independence are not irreproachably extensible to foundational sciences.

The interpretation of $\supset$ through "it is not true that p is true and q is false" comes from Frege.

[On the rule of implication, *Principles* § 38, definition of mathematics and logic *Principles* § 1, § 10.]

Russell. American Journal of Mathematics 28. The theory of implication.

21

Basic concepts $\supset, \sim$, seven formal axioms.

Prothotetik is treated in $*7$, with two additional formal axioms (and the basic concept (x)) and the possibility of the definition of negation $\sim p = p \supset (s)s$ expressed. Beyond this the formal rules *Imp*, *Subst*, and the rule of variable binding.

Definitions have nothing to do with the matter itself, but just with notation.

Whitehead. On cardinal numbers, *American Journal* 24

His definition of multiplication and extension of combinations to infinitely many (binomial coefficients and the binomial theorem). Definition of sum and product for infinitely many summands and factors.

Chwistek. International Congress Bologna

Pure type theory leads together with semantics to the Richard antinomy. All functions that occur in his system build a type ω that shall function as a foundation for the continuum, and all sets given through it are probably measurable.

Semantics and mathematics in one system.

Gergonne. Introduces the name implicit definition.

22

Fraenkel Bologna. A set is called finite if its power set is not reflexive. This definition appears to fall between the ones of Dedekind and Russell. The principle of choice for sets of *finite* sets follows from the ordering theorem. The ordering theorem probably between this proposition and the general principle of choice.

Tarski. Über endliche Mengen. Definitions

1. The Russellian one by induction
2. Sierpinski, smallest class K that has the property:

   a.) $0 \,\varepsilon\, K$ b.) $(x) \,\varepsilon\, K$ c.) $x, y \,\varepsilon\, K \supset x + y \,\varepsilon\, K$
3. Kuratowski. A is finite if $P(A)$ is the only set X for which:

   $X \subset P(A),\ 0 \,\varepsilon\, X,\ (x) \,\varepsilon\, X$ whenever $x \,\varepsilon\, X$

⟦Written at left of 1–3 : Equivalent without the axiom of choice⟧

4. Dedekind's is the oldest definition of finite sets

5.) A is finite if $P(P(A))$ is not reflexive $\equiv 1, 2, 3$

6.) A is finite if it can be doubly well-ordered, P. Stäkel $\equiv 1, 2, 3, 5$

7.) Weber, Kürschak. A first element with each ordering, and if there is an ordering with the set $\equiv$

23

Frege receives a letter from Russell, about antinomies, 1903

Fraenkel Math. Zeitschrift 22, about Skolem's paradox

Poincaré Gedanken. p. 108, p. 134, about all concepts being denumerable

Carnap claim. Impredicative comes from Russell (not semantic)

Heterological from Grelling (semantic)

Antinomy of Richard treated in *Principia,* vol. I, p. 61

Fraenkel, Mengenlehre^D p. 214

Nicod Camb. 19^D ⟦Formulas about the Sheffer stroke⟧

24

Veblen. About categoricity, *Bulletin of the American Mathematical Society,* vol. 12, p. 303

A book description.[7] Various grades of categoricity are investigated here, based on in how many kinds they can be ordered in a one-to-one way. For example real numbers with addition ∞^1, plain order ∞^∞, addition and multiplication 1.

Tarski ⟦Continues the list of definitions of a finite set.⟧

25

Skolem Skrifter 1929^D. (Cites conference Helsingfors 1922)

§ 2 Proof of set-theoretical relativism in which an axiom system is put up. Gives all of the construction principles for sets, taken to be basic operations, axioms are general propositions. (Löwenheim's theorem superfluous.)

§ 1 Enumeration of the possible properties (therefore also sets) in the Fraenkelian as well as Skolemian separation axiom.

§ 3 Set theory in which both of $\varepsilon, \overline{\varepsilon}$ can be false.

[7] ⟦Huntington's *The Continuum as a Type of Order.*⟧

Basic operations $\{x\}, \overline{x}$ (complement), $\dot{x}$ set of sets from x that do not contain themselves, 0 (0-set), this system proved to be free from contradiction (axiom of choice missing).

Further, a rudimentary set theory in which only single ε-members with conjunction, all, there is, and negation are used for separation. Whether free from contradiction is questionable.

§4 Proof of the Löwenheim theorem without the axiom of choice (if at all satisfiable in a domain $\mathfrak{B}$, then already in the domain of the natural numbers). The system of solution of level n (for each n) is ordered so that the ordering is maintained in the continuation.

"Generalized Löwenheim theorem" = theorem for infinitely many counting expressions.

General conception of theories: only universal propositions and all production principles as operations.

26

Such a theorem proved when a contradiction from its negation in the Skolem Löwenheim way.

§5 Provable: If to a theory T of this kind, free from contradiction, axioms of the form

$$\overline{V(a)} \vee V(t) \qquad \overline{V(x\,f(x))} \vee V(xy)$$

are added, it (T') remains free from contradiction. From a solution of level r for T that has a continuation to $2r$ follows a solution of level r for T'. It follows from this that a problem can become decidable through representation axioms ⟦Vertreteraxiome⟧ only if it wasn't decidable earlier.

An indirect proof of existence for $\Sigma_x f(x)$ (free variables occur in x) delivers finitely many individual symbols [built through the basic functions of the theory, starting from constants] for x. The same is claimed for $\Sigma_x \Pi_y A(xy)$. Here, though, the variables possibly built with a newly introduced *Ind* function $y = f(x)$. {In arithmetic moreover through the application of complete induction, a non-finitary production principle}.

§ 7 Example of a domain that is not isomorphic to the number sequence even if it is an integral domain and even if for every two relatively prime h, k, $ah - bk = 1$.

Conjecture that the number sequence is not at all characterisable by counting propositions.

27

Ordinal numbers of the second class only denumerably many, those characterisable through expressions with numerical functions as apparent variables denumerably many, in which such occur again as variables, etc. "There is no possibility to introduce things nondenumerable as anything else but a pure dogma."[8]

§ 6 Some indications that a theory remains free from contradiction if one adds "abstraction axioms." These are axioms that claim the existence of sets on the basis of the separation axiom. 1. If prefixes refer to individuals = abstraction of the first kind. ⟦2.⟧ If they refer to functions = abstraction of the second kind. Hopes to make it clear with this that one can prove the impredicative processes of analysis to be free from contradiction.

Frege Grundgesetze[D] I $\vdash$ assertion sign

Basic propositions:

1. $\begin{array}{l} -\!\!\!\top\!\!-\, y \\ \;\;\;\llcorner\!-\, x \end{array}$ implication
2. $-\, x$ x is the true
3. $x = y$ x ident y
4. $\overline{\top}\, x$ x is not the true
5. $\backslash x$ = the element of x if x is a class with one element
= x in all other cases

1–5 = functions of objects

28

6. $\vdash\!\overset{\mathfrak{a}}{\smile}\!- f(\mathfrak{a})$ $f(\mathfrak{a})$ is the true for each $\mathfrak{a}$
7. $\grave{\varepsilon} f(\varepsilon)$ the course-of-values of the function $f(\xi)$

---

[8] ⟦A paraphrase of Skolem.⟧

6–7 functions of one-place functions
the result an object

8. $\vdash\!\overset{\beta}{\smile}\!— M_{\beta} f(\beta)$

function of one-place functions the argument of which is a one-place function of level one
the result an object

9. Variables

a.)	$a\ b\ c\ d \ldots$	free		of objects
	$\mathfrak{a}\ \mathfrak{b}\ \mathfrak{c}\ \mathfrak{d} \ldots$	bound through	$\smile$	
	$\varepsilon\ \alpha\ \beta$	bound through	$'$	
b.)	$f\ g\ \ldots$	free		functions
	$f\ g\ \ldots$	bound through	$\smile$	

29

Defined: names of objects $\mathfrak{a}$

names of functions, one- and two-place of first level

Rule for these definitions given in a precise way § 33. Through definitions is created as little as in geography when the Yellow Sea is mentioned, ⟦*Grundgesetze*⟧ *I*, p. XIII.

Axioms 3 propositional

2 × Aristotelian

1 equality

1 for $\backslash x$

The one that leads to a contradiction:

$$\vdash (\acute{\varepsilon} f(\varepsilon) = \acute{\alpha} g(\alpha)) = (—\overset{\mathfrak{a}}{\smile}— f(\mathfrak{a}) = g(\mathfrak{a}))$$

Six rules of inference for the propositional calculus (among them, e.g., syllogism, exchange and conjoining of premisses, etc).

One ⟦rule of⟧ setting ahead of the all-sign (in the Polish form, ⟦from⟧ $A \supset f(x)$ ⟦to⟧ $A \supset (x)f(x)$.

One rule of substitution formulated in a precise way for function- and object-variables

One renaming of apparent variables.

Definition of correctly formed names (meaningful names), vol. I, §28–30, p. 45. Even the rule for designating different bound variables is put into effect.

30

The aim of the "Grundgesetze" is to prove that arithmetic is a part of logic – therefore the precise analysis of ways of inference. Page 3: An object can never be a concept, and a concept under which an object falls must not be mistaken for it.[9] Type theory! Contrary to Schröder and Dedekind, a distinction between sense and denotation.[10] "German and Latin letters don't mean an object but indicate it."[11] A function of level one and a function of level two are very clearly distinguished, §21, p. 36 ff. A sign for an object, a sign for a function.

"The expression for a function is amenable of completion, unsaturated." Logic can give an answer to the question why we recognize a logical law as true only by reducing it to another one. Otherwise it will be owing an answer.

Vol. I contents: Theorem about chains (a series that is produced by a relation). Definition of finite and infinite and equivalence with the possibility to be ordered in a series that has a first and a last member. 0, 1, successor, uniqueness, the successor relation, $num[0 \ldots n] = n + 1$, number of.

Vol. II. The intention is to introduce the real numbers. 1. It is shown of certain relations between classes of integers that they do not satisfy the axioms for magnitudes (are positive classes), and real numbers are comparisons between these magnitudes.

Carried through: Definition of "class of magnitudes" = class of relations (composition = addition).

In chapter two, doctrine of magnitudes.

---

9 ⟦Frege almost *verbatim*.⟧

10 Über Sinn und Bedeutung, Zs. f. Phil. u. phil. Krit 100, p. 25.[D]

11 ⟦Frege, p. 31⟧

Vol. II, continuation of cardinal numbers. A part of a finite number is finite. Preparation for the definition of addition.

§ 93 Idea of metamathematics clearly formulated, though only its possibility, examples.

31

Skolem Skrifter 1920

§ 1 Löwenheim's theorem holds also for propositions that arise from counting propositions by finitely iterated application of the formation of sums and products to denumerably many ones. Further *claimed* for all those that are built up by finitely iterated application of $\vee \cdot \overline{} \; (\;) \; \exists$ and the formation of chains (R^*). Further extended to propositions with infinitely many $\exists$ after finitely many.

§ 2 An axiom system for the *calculus of groups* (*Algebra der Logik* I, p. 628) is given. A decision method for universal propositions is specified and it is shown that no new universal propositions become provable through existence axioms.

§ 3 Incidence axioms of the plane with the basic concepts $a = a$, *a lies on* A[12] (without the Desargues property). It has the same property ⟦as in § 2⟧. Again, an existence axiom delivers nothing new and universal propositions are decidable (whether they follow from the axioms). (The axioms are in essence: Exactly one line [point, respectively] goes through two points [two lines, respectively].

The axioms have in § 2 and § 3 always the form: From certain pairs, triples, etc follows this or that etc triple. Therefore it is sufficient to search for the set that is closed with respect to these operations.

32

Frege continued.

Definition of a positival class:[13]

1. Domain and codomain of all the relations are equal between themselves.

---

[12] ⟦Skolem writes *ab* and *AB* for the relations of point and line equality and $a = A$ for incidence of a point on a line.⟧

[13] ⟦The German *Positivalklasse* was invented by Frege.⟧

2. That which is composed of two relations belongs again there.

3. The composition of a relation and the inverse of another one is necessarily the 0-relation and the inversion of a relation of the class.

4. Each relation is one-to-one.

5. The relation of identity does not belong there, $\equiv$ a set of permutations of a class that contains for *none of it the inverse,* and for *each two the composed one,* and for each two either $R|S^{-1}$ or $S|R^{-1}$, whenever $R \neq S$.

A positival class if in addition: to each a smaller one, to each nonempty one with a nonempty complement an upper limit.

From this, a derivation of the Archimedean axiom and the commutativity of addition and the theorem about smaller and greater.

Schröder, calculus of relatives

Starts from the function calculus and the propositional calculus (no axiomatics of its own for the real operations). In addition to dualization also a second principle: Conjugation consists in rearranging formulas, $x + y$, $x\,;y$ replaced by $y + x$, $y\,;x$, all the rest left the same.

33 ⟦A page about resolution of equations in Schröder's calculus.⟧

34

Löwenheim, Annalen 76

1. It is not the case that each expression can be condensed (against Schröder III, p. 551).

2. It is not the case that each sum or product over relatives can be evaluated, because one can express by such that the domain is denumerable.

3. There are no *Fluchtgleichungen*[14] at hand in the calculus of classes (also with =) that would be satisfiable in each finite domain but not in a denumerable one (also if classes occur as apparent variables).

4. Sums and products can always be evaluated in the calculus of classes (also if = occurs), contrary to 2.

[14] Löwenheim, p. 448, explains this as: an equation that is not satisfied for all, but just for some individuals.

⟦Added to the left of 3 and 4: more precisely in Behmann⟧

5. Each equation in relatives is equivalent to a binary one (does this hold also for level 2 **?**).

Schönfinkel, Annalen 92

⟦A list of the basic functions of Schönfinkel's combinatory logic, followed on page **35** by:⟧

Each proposition of the logical calculus can be expressed by the functions introduced so far. Each logical proposition becomes a combination of U, C, S, and one can even avoid the parentheses – or one can do so that U stands in the end and then leave it out.

Axiom systems for the calculus of classes

1. Schröder [$\subseteq$ the only basic concept]

⟦Schröder's axioms for the calculus of classes follow.⟧

36

Müller, Eugen ⟦Replaces three of Schröder's axioms by others.⟧

Huntington Trans Am V Set of ind. post.

[Always a class K as a basic concept, and a law by which $a + b \, \varepsilon \, K$ etc.] Further identity assumed in a contentful way.

I $+ \cdot$ as in Whitehead

⟦Axioms for sum and product follow.⟧

Definition. $a \subseteq b \equiv a + b = b$

II $\subseteq$ as in Schröder

⟦Axioms for the inclusion relation, followed on page **37** by:⟧

The question is posed whether the distributive axiom can be replaced by the uniqueness of negation. Independence is shown, but one needs to require existence only for $a + b$ or ab.

III + ⟦Axioms for sum, followed by:⟧

Interpretations (logical field)

Each *finite* interpretation consists of 2^n elements, and with equal n, each two are isomorphic.

⟦The upper right corner has an addition that mentions a work by Bergmann in *Monatshefte* 36 with "sharper inversion axioms."⟧

38

Menger's field axioms ⟦These are named, source Jahrb. 37.⟧

Sheffer^D ⟦Axioms for the Sheffer stroke⟧

Bernstein B. A. Trans. Am. 28

Claims that the concepts $\vee$ etc are used also contentfully in the propositional calculus, and that one must therefore distinguish between these two uses. He says further that one must have a sublogic that lies at the basis of all theories, and that the axiomatic theory of the propositional calculus is not any more fundamental than any other axiomatic theory. (Therefore also proofs of independence, contrary to Russell *Principia* I, p. 94 ⟦correct page is 91⟧.)

He gives for each axiom system a "complete Exist Theor" (after Moore). The following, for example, is completely independent: 1. $a|a \neq a$ 2. $a|b = b|a$ 3. Exactly two elements (further $a|b \,\varepsilon\, K$).

The aim of the work is especially to give independent axiom systems for the *contentful* propositional calculus. (The addition to the axiom system of the propositional calculus of the axiom by which there exist two elements gives a non-independent one.)

39

In *Bull. Am* 35, the systems are modified so that they become irredundant in the sense of Church.

Trans 17 $\;|$ = basic concept, one postulate less than Sheffer.

Bull 30 An arithmetic interpretation of the propositional calculus (in the field *mod* 2, $a \times b = ab + a + b$).

Bull 30 On a generalization of the syllogism = general transitive Boolean operations.

Bull 37 The Russellian theory is made into a mathematical science (a contentful theory).

Russell is reproached of not having distinguished between the concepts that belong to a theory and those that lie outside it. Propositions state something about the concepts of a theory, and therefore they must contain concepts that are not in the theory. In derivations, concepts must be used that do not occur in the theory.

Church, Trans 27, irredundant postulates

$=_{Df}$ no postulate can be replaced by a weaker one. Independent = no two have a non-tautological consequence = mutually prime after H. M. Sheffer Bull 22 p. 287 (report of a talk).

Completely independent and irreducible are mutually exclusive (Sheffer claims erroneously the contrary).

40

With "completely independent," the domain of mathematical systems gets divided into the greatest possible number of subdomains (2^n), with irreducible ones into the smallest possible number $n+1$.

Necessary and sufficient condition: the negations of each two postulates contradict each other. (If no axiom is a tautology and disjoint contents, then independent.)

From $A, B, C, D,$

$$\underbrace{A \quad \overline{A} \vee B \quad \overline{A} \vee \overline{B} \vee C \quad \overline{A} \vee \overline{B} \vee \overline{C} \vee D}_{\text{have disjoint contents}}$$

If A, B, C, D independent, then irreducible between any two.

As an example, a categorical and irreducible axiom system for + and − for the integers.

Dines Complete exist Theory *Boll* 21, p. 18

Bernstein, Trans 26

Let the operations that form a group be $Axy + A'xy' + D'x'y + D$

$D \subseteq A$

Abelian group $D = A$

Taylor, Annals 19. Complete exist of Bern

Here also a modification of Bernstein by which completely independent.

41

H.M. *Gehmann, Bull* 32. Completion to irredundant. To be able to use the method of Church, the postulates that are maintained have to be only "orderly independent." Then independent and irreducible.

Gegalkine, Rec. math. 34, p. 26

A mechanical method for deciding on truth and falsity in the propositional calculus.

$p+q$ defined: exactly one of the propositions p, q is true. Then all laws for the field *mod* 2.

Rec. 35, p. 369:

"Each proposition with just one-place predicates is provable" proved in Herbrand's way.

Arithmetization extended to propositions with apparent variables.

Rec. 36, p. 332:

The result of Behmann on propositions in which there occur only one-place predicates is derived in a very comprehensive work (125 pages).

Wernick, Journ. f. Math 161

Peano's proof of the independence of the second distributive law through convex point sets. $\cap$ = intersection, $\cup$ = the smallest covering set that is convex.

42

Formal system for the calculus of classes without the second distributive law. ⟦Axioms and rules of the system are given.⟧

Each proposition has the form $V \rightarrow W$

I $a(b+c) \rightarrow ab+ac$ is not provable

II Decision procedure for when a formula is derivable from the axioms

a.) The usual criterion (o, 1) for expression $A \rightarrow a$ or $a \rightarrow A$

b.) A product is implied if and only if each factor is, a sum implied if and only if each summand.

c.) A product implies a sum if and only if either already a summand or already a factor.

New axioms that are added ⟦the axioms follow⟧.

Now also distributivity independent (decision problem remains unresolved).

Distributive law added to the axioms. Then those formulas provable that are correct in the usual sense (as is shown).

43

The same if even the negation axioms are added.

Dubislaw, elementary proof of the freedom from contradiction with a 3-valued table (ambiguous).

Frink, Bull 34, p. 369 ⟦the correct page number is 329⟧

Operations pairs in a Boolean algebra that forms a commutative, associative, and distributive algebra (with respect to addition a commutative group).

⟦Formulas about addition and multiplication⟧

If the Abelian algebra is finite, the corresponding associative algebra is linear over the field *mod* 2 (this representable through finitely many basis elements).

[The algebra is always a direct sum of a 0-algebra and one all the elements of which are idempotent.]

Bell, Trans 29 (weak)

Analogy between arithmetic and Boolean algebra treated, also a concept of congruence is introduced and (divisibility = $\subseteq$) addition, multiplication simply taken over.

Hurwitz, Trans 30. Gives a better definition of congruence that allows to include more properties ⟦these indicated by equations⟧.

44

Solvability condition and resolution of the congruence $ax \equiv b(u)$ analogously to the rational one.

Klein, Jahresb. 38

A distributive system (originator of the definition is König, *Neue Grundlagen der Logik und Arithmetik*), Comm, Assoc, tot lin, distributive from both sides, extended through the requirement that a negation exists as well as a division in two classes (true, false)

Padoa Congress Bologna

Stamm Beitrag zur Algebra der Logik. *M.H.* 22, p. 137

An axiom system with the basic concepts of "incompatibility" and "both false."

2. Pairs (a, b) introduced, and operations are so defined that each linear equation (with coefficients of *the original* domain) has a solution. ⟦Added: the rules of calculation agree with the usual ones for pairs (a, a) ⟧, but the associative law does not hold (not even addition) and the equation has no unique solution).

45

Wiener, Trans Bd VIII, p. 17 ⟦Wiener (1917)⟧

The most general operations that satisfy the Huntington postulate ⟦A formal development follows.⟧

Frink, Trans 27 ⟦Frink (1925)⟧ puts up the above condition ⟦a formal development follows to page **46**⟧

Post, Am Journ. 43^D

A fairly clear distinction between mathematics and metamathematics, declares expressly that all auxiliary means are allowed in metamathematical considerations (only with consistency proofs not).

⟦The following points I–III are cancelled:⟧

I A proof that each truth function can be defined from $\vee, \sim$.

II Each formula that is always correct is provable from the axioms. (A proof in which brought to normal form.)

III A proof of freedom from contradiction from: If one adds one unprovable proposition, then each proposition becomes provable.

⟦Point IV has been changed into:⟧

[IVa General characterization of systems:

I $p\, q \ldots \varepsilon\, \mathfrak{S} \supset f(p, q \ldots)\, \varepsilon\, \mathfrak{S}$

II Rule of substitution

III From formulas of "a certain shape" follow others of "a certain shape," rule of implication.

IV Finitely many formal axioms.]

V ⟦Cancelled similarly to I–III:

Definition. Free from contradiction: If not each formula derivable.⟧

Definition. Closed if each function either provable or through addition a contradiction.

Necessary and sufficient condition for the truth of a function in a system:

Each function of one variable obtained through substitution is true.

47

A function is called completely false if it is false under each substitution (i.e., in contradiction with the axioms).

Completely closed = closed and contains a completely false function.

Necessary and sufficient for true in such a system is that each function that arises through the substitution of a true or completely false function of one argument p is true.

VI Connection between the treatment through postulates and one through truth tables.

Frege antinomies [Über Sinn und Bedeutung^{D}, cf. later]

Ways out:

1. Classes are improper objects. They cannot occur all over as arguments just because they are objects. There are, then, different kinds of improper objects, "relations," "classes of classes," etc. Rule about

which objects as arguments is too complex, therefore this way (type theory) is not viable.

2. Classes are only apparently names, and even some arguments: How could the number of a class be spoken about **?**

A derivation is given by which there exist for each function of the second level (a function of a function of one argument, for example $\acute{\varepsilon}\,\varphi\,\varepsilon$) two concepts that, as arguments of this function, return the same value, without these concepts having the same range.

The axiom that leads to a contradiction is replaced by:

$$\acute{\varepsilon}\varphi(\varepsilon) = \acute{\alpha}g(\alpha) = \;\overset{\mathfrak{a}}{\smile}\begin{array}{l} f(\mathfrak{a}) = g(\mathfrak{a}) \\ \mathfrak{a} = \acute{\varepsilon}f(\varepsilon) \\ \mathfrak{a} = \acute{\alpha}g(\alpha) \end{array}$$

Post, continuation. A condition is also put up for when a system is a "truth system," i.e., when the ground functions can be so interpreted with two values that a proposition becomes provable whenever there results + for each substitution of truth values.

48

VI Many-valued logic   Truth values $t_1\, t_2 \ldots t_m$

$\sim t_i = t_{i+1} \quad t_i \vee t_k = max(i,k)$

It is shown that all functions can be defined from these two. For the system to be completely closed, one has to assume as true only values of a certain subset of $t_1 \ldots t_m$, for example $t_1 \ldots t_\mu$, $1 < \mu < m$. It is indicated how one obtains an axiom system to this effect, and that this is completely closed.

⟦The new version of IV, given above, is inserted here.⟧

[equivalent = mutually contained]

A system that is contained in another has a representation in the other if the functions in the other are defined so that *the same* theorems are provable for these functions. ${}_\mu T_m$ and ${}_{\mu'} T'_m$ equivalent only if $\mu = \mu'$, $m = m'$.

Some conditions are given (though not necessary and sufficient ones), for containment of ${}_{\mu}T_m$ and ${}_{\mu'}T'_m$.

VII Interpretation of many-valued logic.

$(p_1 p_2 \ldots p_{m-1})$ in which $(- - - + + + +)$

$(p_1 p_2 \ldots p_{m-1}) \vee (q_1 \ldots q_{m-1}) = (p_1 \vee q_1 \ldots p_m \vee q_m)$

Truth value τ_k ⟦cancelled: if only $p_1 \ldots p_{m-k}$ true $p_{m-k+1} \ldots p_{m-1}$ false⟧

49

$p_1 \ldots p_{k-1}$ true and the last false.

Hertz, *Ann* 101[D]

Only sentences of the form $a_1 a_2 \ldots a_n \rightarrow b$

Rules of inference:

1. Transitivity in its most general form:

$$\frac{\begin{array}{l} a_1 a_2 \rightarrow b_1 \\ a'_1 a'_2 \rightarrow b_2 \\ b_1 b_2 u_1 u_2 \rightarrow c \end{array}}{a_1 a_2 a'_1 a'_2 u_1 u_2 \rightarrow c}$$

2. From $\mathfrak{A} \rightarrow c$

$$\mathfrak{A}\, b \rightarrow c$$

Initial sentences $\mathfrak{a} \rightarrow \mathfrak{a}$

There exist always independent axiom systems for *finite systems,* for *infinite questionable* [counterexample when only sentences of the form $a \rightarrow b$ are allowed (with a corresponding modification).

Finite systems also decision procedure whether a sentence follows or not.

A sentence of a system is called "inabundant" if no sentence with a smaller antecedent belongs to the system. Sentences $a \rightarrow b$ are "tautological."[15]

If a system contains more than one independent system of inabun-

[15] ⟦Possibly $a \rightarrow a$ is meant.⟧

dant axioms, it contains a circular ordered set of non-tautological inabundant sentences of the form

$$E_1 e_2 \to e_1 \; E_2 e_3 \to e_2 \quad E_n e_1 \to e_n$$

50

This theorem is extended under certain restrictive conditions to sentence systems of sentences of the form

$$f(x_1 x_2 \ldots x_n) g(x_{1_1} \ldots x_{i_n}) \to h(x_{j_1} \ldots x_{j_n})$$

Here the rules of inference are extended by the renaming of variables and by naming two of these the same.

Huntington. The term *complete* stems form him, for what *Veblen* has called *categorical* (*Trans* 5, p. 346) ⟦Added note: Here especially the possibility is put under consideration that not derivable by finitely many syllogisms (but nevertheless objectively determined).⟧

In *Trans* 3, he specifies complete axiom systems for the theories of positive integers, rationals, fractions, and real magnitudes.

Trans 6, p. 17 (footnote) A claim categorical = either P or $\overline{P}$ deducible.

p. 210 (footnote) This limited, "perhaps logic is not complete," after a suggestion by H. N. Davis.

Wilson, Bull 14 (very weak, mainly about the axiom of choice).

States the equivalence of "categorical" and "decidable" depends on the completeness of logic and that one must formalize the logical axioms and rules of of inference for a question of deducibility to have a sense.

51

Wiener, Proc. London 19.

Modification of the Whitehead-Russell theory of measure, for the case of psychology in which a greatest intensity and finite decidability are questionable.

Russell Principia

(Predicative) negative numbers = relations between a and $a - m$ (a and $a + m$)

u/r = relation that obtains between two relations R, S when $R^r = S^u$

$-u/r = u/v|Cnv$

real numbers = sections of rationals, with + numbers as sections of + and − numbers as sections of −

⟦Notions about vector families from *Principia III*, part VI.B, to the end of page **52**⟧

53

Mult ax = there exists for each class of nonempty classes etc

NC ind = the $(+_c 1)$ chain of 0 ⟦cardinal numbers are an inductive class⟧

Infin ax = $\alpha \,\varepsilon\, NC\ ind \supset \alpha \neq \Lambda$ ⟦Added: Same in meaning with this: There exists for each inductive class an element not contained in it. The class of individuals is not inductive.⟧

Sharper: The class of individuals in infinite, $(\geqq \aleph_0)$.

This has typical ambiguity. It is also allowed to have a symbol with different types within the same proposition.

Vol. II, p. XI states clearly that one cannot talk about all types.

A proposition in which no types are yet determined is called a symbolic form. Only symbolic forms have typical ambiguity, propositions or propositional functions instead not.

Ambiguity about type can occur only within the process of "giving sense."

To assert a symbolic form = to assert all the propositions that come out of it through the determination of types.

If one raises *simultaneously* all of the types that occur in an expression, the truth value is maintained, but not if only in part, e.g.,

$Nc\ Ke'Ind \cap \tau'Ind = \Lambda$

$Nc\ Ke'Ind \cap \tau'Ke'Ind \neq \Lambda$

Nc is a doubly ambiguous relation with regard to types.

Permanent truth if proved for the lowest type and all individuals altogether meaningful.

Stable truth if proved for the type and for each higher one for the meaningful ones.

54

Yule, Ann 95 Axiom system for the calculus of classes with the concepts

$a!b, \quad a\text{i}b$ (naturally only one or the other)

$ab \neq \Lambda, ab = \Lambda$

concepts needed. Independent system with four axioms (and as fifth the existence of two elements).

Zaremba, Enseign. 18

Claims for example that, p. 13, $a_x \supset b_x$ with a_x always false is neither true nor false, but makes the convention that it should be true, after all. Presents the going over between contentful and formal proofs very well and comprehensibly. (Proof of a conditional theorem and *Red ad abs* through the addition of a hypothesis, rules of substitution and implication formulated in a precise way.)

Then enrages about "compatibility and independence."

Contentful axiomatic theories pretty clearly presented. Difference between technical terms and current terms: axiomatic basic concepts and concepts of the ground discipline.

Consequence is defined as in Carnap. Indicates clearly the difference between objective consequence and derivability and remarks that the concepts have to be made precise. Method for proofs of contradiction and independence presented in the abstract (interpretations).

55

Tarski. Axiom of choice, *Fund.* V

Propositions of equal value, m, w, y, q transfinite

$m.n = m + w \quad m = m^2$

$$\text{Monotony}\begin{cases} m^2 = n^2 \rightarrow m = n \\ m < n, y < q \rightarrow m + (x)y < w + (x)q \\ m + (x)y < w + (x)y \rightarrow m < w \end{cases}$$

Hartogs, Ann 76 (Sierpinski)

There exists for each power an $\aleph$ that is neither greater nor smaller.

H. Grassmann, Lehrbuch der Arithmetik 1861

Here a definition of addition and multiplication through recursion. (The book apparently takes that not as a definition but as a "basic law.")

Chwistek, Hyp. Mengenlehre, *Zs.* 25^D

1. transcendent = internally comparable, finite, not inductive.

2. prime set = not inductive, each subset equipotent with either an inductive or with the complement of an inductive one.

Transcax = each subset either inductive or the complement of an inductive and non-inductive one. ⟦Changed from: each subset either finite or the complement of a finite or not finite one.⟧

From this follows 1 and 2.

56

It is assumed that the set of individuals is prime, its cardinality Ω. Then a proof of internal comparability, and each cardinality ⟦added: of a subset of the cardinality Ω^n⟧ representable in the form

$\Sigma_0^m \alpha_i \Omega^i - \Sigma_0^p \beta_i \Omega^i \quad \alpha_i, \beta_i, m, p$ inductive

The rule of computation as for polynomials. Gives a non-Archimedean group when extended by fractions.

Indication of a proof of freedom from contradiction for Transcax, one considers only inductive and complements of inductive sets of individuals as (proper) sets and tries to define also for higher types "proper" sets in such a way that the closure with respect to the axiom of choice is maintained. By this, obviously, independence of trichotomy (axiom of choice from the rest of the axioms of set theory).

Problem. Is it possible to add without contradiction to Transascax ⟦An addition at left states: Affine axioms⟧:

a.) In case α, β internally comparable and comparable within the two, then also β comparable with $\mathfrak{A}'\alpha$.

b.) From a well-ordering for α follows one for $\mathfrak{A}'\alpha$.

Zs. 14

1. If one is to maintain the rules of the algebra of logic, there remains nothing else for avoiding the contradictions than a limitation of the domain of arguments, therefore type theory.

2. Richard's antinomy.

3. A lot of mathematics possible without the reducibility axiom.

Lebesgue measure in *Lemberg Ak. Wiss.*

57

Peano. On mathematical definitions. Bibl. Congr. Paris^D

Law of homogeneity: The same free variables must occur at left and at right and in the same way. (For example, it is incorrect to define $\frac{a}{b} + \frac{c}{d} = \frac{a+c}{b+d}$, because rational numbers at left.)
Each definition is an abbreviation and therefore in principle superfluous.

Question: Can one define something in a false shape? One has to ask whether one can define something from certain given concepts.

Fractions as pairs, Stolz, allg. Arithm p. 43 1885

Tannery Leç d'Arithm., 1894, p. 48

Fractions as operations, Méray, Leçon sur l'An. infin, p. 2, 1894

Already in the Rhind Papyrus

Padoa Bibl. Cong. Paris^D

The basic concepts of a theory are said to be independent in relation to the axioms of the theory if, for no (a), an equation of the form

$a = [\quad]$ is provable from the axioms.

Proof method for independence is to give for the basic concepts interpretations that satisfy the axioms, and that satisfy these axioms also when one changes the interpretation of *one* concept and leaves the others the same.

58

There follows, after an introduction to the content of axiomatic theories, an axiom system for whole (+−) numbers with the basic concepts "number," "+1," "x," designated by *sym* x. This, though, is not complete because the residue class, for example, satisfies it.

Burali-Forti defines in Bibl. Con. Paris^{D} the integer, rational, and real numbers as operations that area applied on two-place operations $x + y$, that satisfy the axioms for magnitudes, and that give as results a one-place operation (the k-fold of).

Sheffer, Nat. rel. 6-th Congr. Phil. 1926^{D}

Problem: When are two axiom systems transformable to each other? (Basic concepts of one definable in the other, and theorems provable, e.g., geometry with different basic concepts.) Invariants sought, they are

Tropicity = class of self-transformations that take basic relations to themselves. (Given as a sufficient condition.)

Straticity claims one can derive each from a "homogeneous" one through the assumption of special elements. Number of these elements = static.

All of it *extremely* vague and imprecise.

59

Principles, Paradoxes and type theory   Russell

p. 7 Variables can designate "completely arbitrary" objects.

p. 1 Mathematics is the set of all propositions of the form $f \supset g$ where there occur only logical constants in f, g. Hangs together with the fact that mathematics deals with relational types (= classes of relations that can be defined by just logical constants. The conception of *a contentful axiomatics* lies at the base of it all. *Symbolic logic = the doctrine of inference.*

Propositional calculus. Basic concept *implication* on (x) (in the way of Peano, in the form $\supset_x$).

Calculus of classes. Basic concept $\varepsilon \quad x\ni(\ )$, propositional function.

{Peano is the first one to clearly distinguish between $\subseteq$ and ε, Frege as well.}

Here it is assumed, already, that implication means nothing else but $\sim p \vee q$.

One cannot separate out the propositional functions as independent things in $\varphi(x)$, otherwise contradiction.

⟦In chapter⟧ X, p. 101, Russell explains how he found his contradiction.

(Examples of concepts that apply to themselves: the negative concepts, not-concepts.)

The contradiction is given in the form: There exists to each class u a (sub)class that does not contain it and is not contained in u.

Proposal for a solution: The "quadratic" function $[\varphi(f\varphi)]$ defines no class of the same type as the individual – the others perhaps. Cf. also § 484.

60

Simple type theory is put up in Appendix B, but still a free range of significance = one type, or a sum of several types.⟦Added: The sum of several types is always also a type, but different from the type of the summands.⟧. The properties and relations of daily life (classes as one) belong to the type of the individuals, higher types only with the functions of symbolic logic.

Two basic laws of type theory:

1. To each proposition belongs a range of significance.

2. Ranges of significance form types, i.e., sums of certain pairwise disjoint classes (minimum types).

If $\varphi(x)$ meaningful and x, y in such a minimum type, then also $\varphi(y)$ meaningful.

All propositions make up a type, he believes even that all classes make up a type and that all numbers make up a type. He acknowledges, though, the difficulty. (So types of infinite order are allowed.)

$x \,\varepsilon\, x$ is sometimes meaningful, namely when x is of infinite order. This form of type theory leads to a new contradiction.

New contradiction: $(p)\, p \,\varepsilon\, m \supset p$ associates to each set of propositions a proposition, and to different different ones. Let w be the set of propositions thus obtained that are $\sim \varepsilon\, m$.

$$q = (p) p \,\varepsilon\, w \supset p \qquad q \,\varepsilon\, w \equiv q \sim \varepsilon\, w$$

He doesn't know a solution – perhaps a hierarchy of propositions.

61

p. 114. Definition by abstraction contains the error that it doesn't deliver the object in a *unique* way. Therefore the proposal of taking "class" as a common property. (The Italian school distinguishes between three kinds of definitions, non, post, abstr, Burali - Congr. Paris.) (About definitions by abstraction, cf. Peano F. 1901 §32 '0 note.)

Difficulties connected with the axioms of choice and infinity are not yet recognized (the two definitions of infinity seen without further ado as equal.) The presence of infinitely many objects is supported on empirical grounds (infinity of space and time).

Ratio = relation between integers

Fraction = relation between magnitudes

Bolzano, Parad §13 } Proof of the axiom of infinity

Dedekind № 66 through mapping of things on ideas

On the whole, Russell saw that there exist infinitely many things, should clearly be quite evident.

Russell, Rev. Met Mor 1906^D, p. 630

States clearly that the justification of principles comes from intuition, but not all principles must be intuitively clear at one, they must instead just lead to intuitive consequences (intuition plays the role of sense perception). No class theory necessitates the axiom of infinity and gives only the types up to ω_n.

62

III Here, for the first time, the vicious circle principle stated in a form with apparent variables. *This principle must result as a consequence of the logical theory.* (Limitation in the range of variables foreseen in the universe of discourse.)

Propositions divided by the number of variables that occur. A general concept of definition is impossible because definitions with arbitrarily many variables exist. It follows from the vicious circle principle that a class cannot be an individual ⟦written above: entité⟧, because it contains always an apparent variable ($\hat{x}\varphi(x)$) – or it doesn't exist at all.

Example where RedAx true, *namely for finite sets.*[16]

Indication that everything extensional in mathematics, therefore the ordering axiom sufficient.

Poincaré, Mor et met 1906[D]

A. 1. Hilbert circular because of the proof of freedom from contradiction.

2. One cannot define the natural numbers through the principle of induction, but they are instead to be defined as the numbers that are obtained through successive addition. The theorem of complete induction does *not* follow from this.

B. The reason why it doesn't follow is that the definition is not predicative. If the definition of a concept N depends on all objects A, it can be afflicted by a circulus vitiosus in case there exist among the objects A ones one doesn't define without the use of the concept N. Therefore he adds with such definitions always: all A except those that one can define only through N.

———

16 ⟦Added citation: Les classes doivent être reg. comme des parties purement verbales (symboliques) de jugements et non comme des partiès des faits exprimé, p. 649.⟧

Axiom of choice will always remain undecided. He himself tends to admit as a synthetic *a priori*.

63

This definition of non-predicative is given under IX as a true solution.

Schröder, Alg. d. Logik I § 23 Inverse operations^D

I § 24 Symmetrically general solutions and II_2 § 51 the problem of general equation with two unknowns solved, but parameters must be exchanged with their negatives (three parameters). ⟦An asterisk points at Whitehead below.⟧

II_1 Inequalities. 1§. Each problem that can be expressed with the help of the operations of the propositional calculus and the calculus of classes can be brought into the form: an equation and a number of inequalities.

Individuals, II_1 § 47 – He saw that postulates are needed for the existence of individuals and for the possibility to decompose classes of individuals.

II_1 § 48 Syllogisms with two premisses and three members (with one middle member), but a notation different from *a e i o*.

II_1 §28 Proposition intended as the class of occasions of its use, validity of these.

Propositions with a fixed meaning: necessary $= 1$, $v = 0$.

Special principle of the propositional calculus $(A = \dot{1}) = \dot{1}$ (the principle of assertion)

Whitehead, § 35–37 Johnson method

To solve an equation with n unknowns symmetrically, with $2^n - 1$ parameters, but symmetry not in a quite strict sense – the same with the method of Whitehead, but not

64

strictly symmetric, and also not when exchange of arguments is allowed.

Schröder, Alg. Log. I, Supplements 4, 5

Calculus with algorithms and groups, independence of the distributive law. ⟦An asterisk indicates a footnote, perhaps the previous one about Whitehead.⟧

Supplement 6, number of "types" of logical functions of n variables (two belong to the same type if each comes out from the negation of the other through an exchange of the letters. Solved by

Clifford for $n = 4$ $z = 398$ (Proc. of Lit and phil soc. of Manch 16, 1877)

On the types of comp. statements involv. four classes.

(= number of choices that don't carry over a single one)

Supplement 6 also proof that the equation $xyz + \overline{x}\overline{y}\overline{z} = 0$ with three parameters is not symmetrically solvable.

Vol. II_1 § 34.

All possible propositions (through just one equation) of n domains $= 2^{2^n}$

All possible propositions with even disjunction and negation of n domains $= 2^{2^{2^n}-1} - 1$

Determined by Peano for $n = 2$, $= 32767$

§ 36. Reduction of all possible propositions of "the second level" to truth functions of equations.

Vol. II_2 § 55. Condition put up for associativity of a connective and solved symmetrically generally.

Further, a condition for the distributivity of two connectives.

65

Proved also by Korselt, Bemerkung zur Alg. der Logik, *Ann* 44. Independence of the distributive law with a geometrical example. Schröder proof without negation, this also with negation.

Royce. Frees himself from the axiom of continuity and proves existence of partial systems that satisfy the axioms of geometry (i.e., the uniqueness of the intersection point). ⟦A couple of formulas follow.⟧

The report of Royce in the *Encyclopäd. der philosophischen Wissenschaften*, Tübingen 1912.[D]

⟦There follow the axioms of Royce.⟧

66

Chwistek ⟦1925⟧. Wishes to put up a system in which no other preconditions than those of the logical calculus. ⟦Added: believes that there is no other possibility of a foundation. For otherwise proof of freedom from contradiction, and carried out precisely with the help of the logical calculus.⟧

He claims that one can prove existential theorems in this system only if one has an example.

[Basic idea of Russell's type theory: the concept of "all objects" is meaningless, and even the concept of all properties of x is meaningless.] Chwistek adds that one can build a system of logic and mathematics free from contradiction in which all properties of x occur, but the paradox of Richard is not excluded.

"pure theory of types". . of *constructive types,"* ⊛ = ramified type theory. He claims the contradiction arises through the axiom of reducibility *together with the Russellian theory of classes and relations*. [p. 14 emphasizes that if one applies the rule of substitution to the definition of $f(\hat{z}(\varphi z))$, one comes to undesired results.] [p. 17. The axiom of reducibility cannot be expressed because of the variable $\varphi \quad \exists\psi!\psi x =_x \varphi x$]

⊛ all functions of a given type are constructible. In a second phase of the theory, axioms are added, *axioms for finite numbers, axioms for the continuum*. These are more general than the axiom of reducibility and of little importance for the theory. Congr. Pays slaves ⟦Chwistek (1930)[D]⟧, main objective Poincaré's rule, never to consider objects other than definable in finitely many words. – This made precise (the opposite axioms of choice and reducibility). Système de métamathematique arises through the addition of semantic concepts and corresponding propositions to those of pure logic, and of "allowed infinity," axiom of choice to be proved,

67

and the continuum hypothesis made superfluous.

Claims that Lebesgue integral definable.

Development of the theory: $\mathfrak{I}(xy)$ of the same type.

Identity of two functions is defined as the sameness of range.

In part I, the following are developed: propositional calculus and properties of $\mathfrak{I}(xy)$, further classes, relations, identity, descriptions. ⟦Added: Chwistek call his foundation "nominalistic."⟧

Intax ⟦axiom of intensionality⟧ which says that two propositional functions are identical only when they are built up in the same way.

The axiom of infinity follows form it.

Axiom of nominalism: There exist in each type only denumerably many functions. Zermelo's axiom follows from this.

p. 110, Schröder-Bernstein theorem (for several types of classes).

Arithmetic of cardinal numbers $(+, ., exp)$.

A certain type of inductive numbers is defined and the group property with respect to *Add, Mult, Exp* shown for this.

J. Royce. A strange theorem: Each element lies between x and $\overline{x}$, if x between ab and between ac. Then also between each element that is between bc. For each original y, y is the first and $\overline{y}$ the last element.

⟦line diagram⟧ An element can often lie at the same time between ac and cb (except for c).

68

If the "basis" is brought over into a negation, the the sum goes over into a product.

The geometrical axioms are not rigorously proved, but just indicated.

Steps ahead against Kempe:

1. The axiom of continuity is put apart (and criticized).

2. The treatment is extended to infinite systems, by which the "continuity" in geometry can be introduced.

The method of *Kempe* is put alongside that of Peano and Russell. Advantage that it calls only for the calculus of classes.

3. The basic relation is symmetric.

⟦A criterion for "flat collection" is given.⟧

Additions to *Kempe* (more precise proofs) in Schröd. II2 supplement 8.

Hölder, Sächs Ak., Vol. 78

It is shown that one can construct the upper limit for a sequence of irrational numbers in a noncircular way.

Cf. Hilbert

Weyl Kontinuum

69

Weyl, Symposion. Phil. Zeitschrift für Forschung und Aussprache, vol. 1^{D}

p. 14 Examples of construction principles that are circular.

Die heutige Erkenntnislage in der Mathematik

Frege, Sinn und Bedeutung. The way of givenness is contained within the sense.

Zs. f. Phil. und phil Krit, vol. 100, p. $26.^{D}$

"Der Gedanke" in *Beiträge zur Philosophie der deutschen Idealismus*^{D}. "Thought" = not real and not psychic.

p. 69. Thoughts are neither representations nor things of the external world. A third realm must be admitted.

(He is, for example, a harsh realist with things of the external world.)

Abstraction principle, "Grundlagen der Arithmetik," pp. 63–68^{D}

70

Becker, Jahrb. Phil phän XI, p. 497

1. Additional axiom $-\sim p \prec \sim\sim p$. A sharper axiom, means that

possibility is an "absolutely" modality.

$$-\sim p \prec \sim\sim p \ = \ \text{possible} \rightarrow \text{necessarily possible}$$

by which six basic modalities:

true, false, impossible, possible, necessary, not necessary

Law of excluded third does not hold for necessary. Modalities have a ranking order.

2. Additional axioms $\sim - p \prec \sim - \sim - p$

$$p \prec \sim\sim p$$

Ten modalities, the new ones are:

impossibly impossible = necessarily possible | negative
possibly necessary | negative

Ranking order: necessary, possibly necessary, true, necessarily possible, possible

Conjecture: Is p or $\sim\sim p$ stronger, cannot be decided by Lewis.

It is shown already from the axiom: $W \rightarrow N^i M$ for each i, a ranking order follows, namely of an order type η (all + modalities can be composed from M and N).

71

Löwenheim, Ann 68

1. Determines the possible subtypes of the group $\mathfrak{G}(a_1 \ldots a_n)$, only the number determined, Schröder I supplement. Each such has as a basis some sum of constituents of the Boolean development, namely so that its sum = 1 (easy enumeration).

Müllerian verification theorem whether an equation, ordered up to summation, is correct. One recognizes this by substituting for the domain 0, 1. ⟦Details and formulas about solutions follow.⟧

Particular solution for equations with arbitrarily many unknowns in a closed shape (law of formation), p. 189 (natural solution). ⟦Added: possibly Whitehead's book⟧

[Not to forget the *mean value theorem* in Schröder.]

States the number and most general form of possible solutions with parameter p, and the possible reprod. solutions (advent? requirement) and number of general solutions. ⟦Addition on next page:⟧ From this, proof that an equation of grade n cannot be solved with less than n parameters.

72

One obtains from solutions again solutions if one combines linearly sets of domains through a disjunction. One obtains each solution of an equation by substituting in all possible ways the values 0, 1 for parameters and by projecting down the solution of the modular equation thus obtained by the parameter constituent that is 1 under this substitution. One solves a modular equation by combining linearly all of the modular solutions with the constituents of the arbitrary parameter (Jevons). By this way, the number of solutions determined.

Löwenheim Ann 73

Mainly results of Whitehead extended to n variables. ⟦Examples of these results.⟧

Weyl, Kontinuum

Circulus vitiosus § 6 p. 19 ff. Proof of Cauchy's convergence principle p. 57, Hilbert's corresponding proof *probably Ann* 88.

73

Löwenheim Ann 79 Gebiete Determ.

Addition *Ann* 68 ⟦An addition to Löwenheim's paper covered on page 71 above. The rest of the page deals with matrices $|a_{i_k}|$ with parameters as elements, and their determinants with the general conclusion:⟧

A system of linear functions can take all combinations of values if and only if Det = 1.

74

Hölder Sächs. Ges. 1901

Axioms for greater $>, +$:

1. $a > b\,.\,o\,.\,a = b\,.\,o\,.\,a < b$ (o = disjunction)
2. $a > b \equiv (\exists c)\, b + c = a \equiv (\exists c)\, c + b = a$
3. $(\exists b)\, b < a$
4. $(a + b) + c = a + (b + c)$
5. $\alpha \neq \Lambda\,.\,-\alpha \neq \Lambda\,.\,x\,\varepsilon\,\alpha\,.\,y\,\varepsilon -\alpha\,.\supset_{xy} x < y$

   $:\supset (\exists c)\, u < c \supset_u u\,\varepsilon\,\alpha\,.\,u > c \supset_u u\,\varepsilon -\alpha$

1–5 → the system of magnitudes is isomorphic with the + of real numbers. There is essentially only one way to introduce a distributive multiplication. Application: Measure of strength (executed precisely).

75

Behmann Ann 86[D] The treatise is written in a contentful way (not axiomatically).

1. Normal form for propositions without *Id* and operations only on individuals.

Only components of the form:

$$(x)\,.\,f(x),\ (x)\,.\,f(x) \vee g(x) \ldots (x)\,\overline{f}(x) \vee g(x)$$
$$(\exists x) f(x)\ \ (\exists x)\, f(x)\,.\,g(x) \ldots$$

[and possibly $f(\mathfrak{a}), p$]

I.e., $\alpha + \beta + \cdots + \overline{\gamma} + \overline{\delta} \cdots = V$

$$\alpha\beta \ldots \overline{\gamma}\overline{\delta} \neq \Lambda$$

Driving in of the operators, this leads also to protothetic ⟦added above: for the solution of the decision problem⟧

I First step in the elimination (no = sign occurs so far)

$$(\exists \varrho)\, \alpha + \beta = 1\,.\,b + \overline{\varrho} = 1\,.\,\gamma\varrho \neq 0\,.\,\gamma'\varrho \neq 0 \ldots \delta\overline{\varrho} \neq 0\,.\,\delta'\overline{\varrho} \neq 0$$

I.e., the elimination from a proposition of the form $F(\varrho\varphi xy\ldots)$. F brought to the above normal form to start with, as:

$$\overline{x} + \varrho = 1\ x\overline{\varrho} \neq 0$$

R: $\alpha + \beta = 1\,.\,\exists uu'vv'\ \gamma(u)\beta(u) \mid \gamma'(u')\beta(u')\quad |u \neq {}^{v}_{v'}$

$$\delta(v)\alpha(v) \mid \delta'(v')\alpha(v') \quad |u' \neq {}^{v}_{v'}$$

Schröder, *Algebra der Logik*[D], p. 121?

76

Intuitively ⟦Diagram, at right the following:⟧

Analogously the elimination of arbitrarily many concepts (either all of them all, or all of them exists).

Applicability to the calculus of relatives, some exercises of summation and formation of products are cleared more simply.

II Identity allowed

Normal form = propositional connectives with elementary components of the form:

$\alpha + \beta + \cdots + \overline{\gamma} + \overline{\delta} \cdots = V$ with the exception of at most n individuals

$\alpha\beta \ldots \overline{\gamma}\overline{\delta}$ consists of at least n individuals

In the final result, there occur only propositions of the form:

1 consists of at least n individuals

0 leaves out at most n individuals

I.e., the second proposition is a propositional combination of propositions of the form: the domain contains at most n (at least n) individuals, therefore, with the exception of finitely many domains, the truth values are uniquely fixed.

To effect the elimination of the class symbol from the normal form, one introduces again operations on individuals, and shifts then these out over a ρ operation. Thereby the ambiguity disappears and reduced to earlier.

77

Formulaire I p. 28[D]

Poincaré, Letzte Gedanken

I Rejection of the non-predicative definitions. Infinite sets cannot be considered as closed, new elements can always become added, therefore a relation to the totality of the elements is meaningless.

Predicative definitions = are maintained under the introduction of new elements.

Non-predicative definitions = are not maintained under the introduction of new elements.

II The foundation of natural numbers on type theory is circular. Poincaré proves that if x, y are numbers of the order n, their sum is of the order $n = 1$, but for this, the concept of "a finite number" is already necessary. (Poincaré doesn't understand the axiom of reducibility.)

III Example in which the Zermelo axioms are free from contradiction (because of predicative concept formation).

IV Only those concepts are allowable that are definable in a finite number of words, and each theorem is, in the end ⟦added: also those that concern the infinite⟧, about the natural numbers, i.e., about what is finite.

[The antinomy of Richard comes about because the association between numbers and laws is not well-defined.]

Poincaré makes strong, conscious objections because he believes that logic and psychology are not independent of each other (p. 142).

78

VI *Mathematics and logic.* The difference between his (Poincaré's) world view and that of the Cantorians and Russellians: Poincaré is idealist (pragmatist), does not recognize the objects of mathematics to exist independent of thought (therefore infinity is only potential, criterion finite definability). Rejection of the non-predicative. A statement must be verifiable and its meaning consists precisely in the possibility of a verification. *Russell* etc are realists, objects are not created by definitions, only their characteristic marks. Mathematics is discovered, not created (Plato). Therefore impredicative possible, and the fruitfulness of the new logic depends on this. (Otherwise it consists of only trivial tautologies.)

Dedekind, Was sind und was sollen die Zahlen

Arithmetic = a part of logic, and this indeed the crucial relation (III, IV)

Each proposition of mathematics = proposition about natural numbers. Second definition of finite (p. XI):

a finite = there is a self-mapping that maps no proper part onto itself.

⟦§⟧64, definition of the infinite, ⟦§⟧66 proof that there exist infinite sets, a similar consideration in § 13.

Paradoxes of the infinite. ⟦§⟧73 explanation of the concept of a natural number = a simply infinite system *N* ordered by φ in which one disregards the constitution of the elements and considers instead in an ideal way the relation in which they are set by φ, so then these elements are called natural numbers (p. 21).

⟦§⟧59, ⟦§⟧60, ⟦§⟧80 Theorem of complete induction, ⟦§⟧126 definition by complete induction.

79

⟦§⟧130. It is shown that that there exist relations φ for which the theorem of induction is applicable but not definition through induction (namely if the series of numbers runs back to itself).

⟦§⟧71. Definition of a simply infinite system = Peano axioms.

⟦§⟧134. isomorphism of all simply infinite ordered systems. This a justification for the definition of the natural numbers.

Addition, Multiplication, exponentiation are developed.

⟦§⟧159 ff. Each finite system has one and only one "number," theorems about these (e.g., number of a sum).

Curry. (1930)

Examples of equations: $BBC = B(BI)$, $BWK = BI = I$

$I = WK$ Definition

B contraction function, *C* exchange function, *K* constant function

W repetition function $Wxy = xyy$

Combinator = combination of B, C, W, K

Combination = somewthing that lets itself be built through "application"

16 combinatorial axioms, 4 combinatorial rules

E.g., $\{\, Cxyz = xzy\}$ for arbitrary xyz

$Kxy = x$

80

1. There exists a combinator K for each combination of variables X from the variables $x_1 \ldots x_n$, such that $Kx_1 \ldots x_n = X$ provable.

2. The same combinator cannot represent two different combinations of variables (in the sense of the same provable from the axioms).

3. If $Kx_1 \ldots x_n = X = K'x_1 \ldots x_n$ follows from the axioms, then also $K = K'$ follows.

B, C, W are conceived as substitution processes of an elementary art from which complicated ones can be composed. Here, by a substitution is understood the operation that arises from any function, through some substitution of the function in an argument place.

Curry (1929) View about a formal theory: Principles for combining the basic things, which produce new things. Subclass of the things produced (assertions). Finitely many axioms which say that certain combinations of the basic things are axioms, and rules, for example $p \vee \sim p$ belongs, for arbitrary p, to the assertions.

Schönfinkel's thoughts are completed by Curry's belief that a particularly simple formal system can be given if, in particular, the substitution rule is replaced by a simpler one.

81

[Russell, definition of cardinal numbers as relations, uninteresting]

Curry (1931) p. 154

Proof of the substitution principle, five new axioms about Π for this.

The theorem to be proved is:

In case $\Pi_n X$ holds, then also $\Pi_m(UX)$ holds.[17]

---

[17] ⟦Added: This proposition is presented formally, *without the use of variables,*

Here U is a regular combinator of the order m and grade n. That is to say, we have:

$$UX_0x_1\dots x_m = \Pi x_0y_1\dots y_n$$

Here y_i combinations of $x_1\dots x_n$.

Couturat, Rev met mor 1906[D]

p. 215 Refutation of the psychological objections (= one needs intuition for discovery).

p. 219 Tous les logiciens sont prêt á rec. que leurs princ. procedènt de l'int. intellect.

p . 221–226 Answer to the argument about circularity of the definition of 2.

a.) In part through the use of word-language instead of symbols.

82

b.) One does not need the concept of number 2 to think about two objects (in a relation). The refutation of the false objection against the definition of numbers and proof by complete induction.

Poincaré, Wissenschaft und Hypothese[D]

Complete induction VI inference The proof by complete induction forces itself upon us with necessity, because it is only an activity of our own understanding.

Mathematical and physical induction have different grounds but their courses run parallel (both proceed from the particular to the general).

Two means in the construction of the implicit, and derivation of the properties of constructions from those of the elements.

Carnap, Constitution ⟦Carnap 1928[D]⟧

Cites Russell (Occam's razor):

> The supreme maxim in scientific philosophising is this: Whenever possible, logical constructions are to be sub-

---

and then it is shown that it is formally provable.⟧

stituted for inferred entities.

To constitute an object c from a means to give a rule by which one can reshape each proposition about c into a proposition about a.

The means for ascending to higher types of objects are just the formation of classes and relations (therefore extensional). About this, p. 57 ff. Justification of the extensionality thesis consists in: The rational propositions about propositional functions are not about it? but about ?? in a "sense," therefore all propositions about propositional functions extensional.

83

Russell. Logical constants don't denote any entities, they can never be logical subjects (are not names). Such express the form of a proposition, not a constituent of a proposition.

Wissen von der Aussenwelt, p. 277 (an empiricist even about mathematical logic)

Russell, Philosophie der Materie

Nagy, Mon H. 5

$T_n^2\ T_n^3\ T_n^4$ given

Frege. Sinn und Bedeutung[D]

The reason for having the sense of a proposition = truth value is that the denotation of a proposition must not change when the denotation of a component changes. Words have their indirect denotation in intensional propositions (to believe, to fear, to rejoice, to expect), i.e., they denote what usually would be their sense.

84

Begriffsschrift 1879 (Peano, first symbolic treatise only in 1889)

6 axioms for the propositional calculus, 2 equations and the function axiom, one rule of inference. On p. VII, the possibility to have derived rules of inference is considered.

History of the antinomies Burali published in 1897.

Greatest cardinal number from Russell in *Principles*, 1903, and letter to Frege, § 344.

P. L. M. S. Nov. 1905. Here impredicative, but still no epistemological ⟦paradox⟧

Richard, Acta math 30 (May 1906)^D, printing of a letter to the publishers of the *Revue gén. des sciences*

König, June 1905 talk in Budapest academy, published *Ann* 71 ⟦61⟧, antinomy of the smallest ordinal number not finitely definable. As concerns the continuum, he misses the Richard antinomy and just concludes that the continuum cannot be well-ordered and that not all elements are definable.

Dixon P. L. M. S. Nov. 1905, König paradox of ordinal numbers and *the smallest integer not definable with a hundred words.*

Poincaré May 1906, Richard

Russell January 1906, number with less than a hundred words Z

85

Autological, Grelling, *Friesische Schule* 1908^D

Frege. Begriff und Gegenstand^D

p. 201 The propositions that can be made about a concept are not at all based on the object. The relation of an object to a concept is different from the similar one of a concept of the first level to one of the second level.

p. 204 In the second, differences of greater importance can be formulated.

Löwenheim, Sys. Phil 21

Introduces calculation with *n*-tuples of domains, in which the algebra of numbers and the calculus of domains ⟦Gebietekalkül⟧ are included.

O. Hahn, Sys. Phil. 21 ⟦15^D⟧

Studies the conditions that must obtain between the coefficients of a logical equation for arbitrary combinations of values for the unknowns $x_1 \ldots x_n$, as long as they remain between the bounds that result for each single one of the unknowns, to satisfy the equation = that the solution can be given through n limit domains ⟦Grenzengebiete⟧.

Tarski, propositional calculus

n-valued propositional calculus $\{0\,\frac{1}{n}\,\frac{2}{n}\ldots\frac{n}{n}\} = \mathfrak{A} = L_{n+1},\ n+1$ values

$\mathfrak{A}^*\ \{1\}$

86

$\sim x = 1 - x$

$x \vee y = min(1, x + y)$

⟦Written at right of the equations: Analogously for $n = \infty$, or $0b$ is called an infinite subset from $[0,1]$ that contains 0, 1, and n-adic? operations.⟧

$L_n \subset L_m \equiv m - 1 \equiv 0\ (n-1)$

$L_{\aleph_0} = \Pi L_{n_i}$ for each growing sequence n_i

In case $n - 1$ a prime number, there exist only two superordinated systems, namely L_2 and S (Tarski).

Each L_n is axiomatizable, for L_3 Wajsberg offprint.

Sufficient criterion for axiomatizability of a system given through a finite matrix is unknown.

Whether $L_{\aleph_0}$ axiomatizable is unknown, Lukasiewicz conjectures:

Complete systems free from contradiction $= 2^{\aleph_0}$ [when also axiomatizable, $= \aleph_0$]

Each L_n has a basis of exactly n elements, in each basis at least three propositional variables, and for L_2, $CpCqCrp$ belongs always.

Carnap, Symp. Proper and improper concepts, I, 1927, p. 355[D]

p. 364 Equivalence of definiteness with respect to decision and monomorphy is claimed. Clearly against implicit definition, to the extent that no specific concept is defined through it of which it could be decided whether it applies in a case. Further, the w-proposition of excluded middle does not hold.

Dörge and Dubislaw. Über Satz von ausgeschlossenen Dritten ? ⟦Announced as appearing in Dubislav (1925), p. 202.⟧

87

Fraenkel p. 253 Sitzb. Berl

An initial set $M = a_1\overline{a_1}a_2\overline{a_2}\ldots$, each constructible set is symmetric with respect to a subset that contains almost all elements of M.

Heiss,[D] Formation of paradoxes with the solution that there possibly exist contradictory propositions that are both true.

Peano, vol. I, contents, 1895[D]

II Theory of real numbers and log and exp. } only simple propositions

III Theory of the integers } only simple propositions

IIII Doctrine of magnitude, *Riv di Math.* III p. 76

V Continuation of real numbers, upper, lower limit, Dedekind property etc, simple

VI Set theory

VII VIII Limits and series and infinite products

IX Theory of algebraic fields

Reone 13, 1905, Boutroux on mathematical functions (that, in his opinion, cannot be described in an exhaustive way in logical systems)

Met et Mor 14, 1906[D] has a matter-of-fact discussion between Russell and Poincaré

Russell Rev met 14[D]

I Role of intuition and the argument that the infinite is not the cause of a contradiction

II no class theory and simple type theory

III vicious circle principle and ramified type theory (in a rather incomplete coverage)

88

Hahn-Neurath (1909) Dualismus in der Logik, *Arch sys.*[D]

⟦Gives a new axiom system for the calculus of classes.⟧

Bolzano 1837 *Wissenschaftslehre*

Hertz, Ann 87. Only propositions of the first grade, i.e., $a \to b$

Sufficient for the uniqueness of an axiom system that there is no cycle. It is indicated, for a special case (simple systems without cycles), how one can construct by ideal elements an axiom system with a minimal number of propositions. And in general, how one can construct axiom systems with a low number of propositions.

89

Chwistek, Zs 30^D

p . 709, 708 $\{\ p \supset p$

an p the result of substituting p for p in p

Through the sign, no distinction between a sign and its denotation.

p. 711 The system of semantics is not extensional.

p. 707 The antinomy of Richard re-enters through a reasonable doctrine of types.

Aim of it all to state the rules so that the problem of mechanization of the aprioristic sciences is solved. Intuitive semantics should be limited as far as possible, and especially made independent of the doctrine of types.

Axioms of the function calculus. 1. Nicod ⟦A rule of inference follows.⟧

90

Russell Principles, p. 397, p. 429, speaks clearly out what is meant by the definition of implication.

Carnap, Logistik

Admissibility of infinite propositional connectives:

against: Hilbert Über das Unendliche, for it Ramsey

Weyl Handbuch 13

Classes as functions: Russell *Einf* 183, Carnap *Aufbau* § 32^D

Type theory in the real sciences Carnap *Aufbau* § 29[D]

Extensionality thesis. Wittgenstein 243 f., and Russell Preface, in addition *PM* p. XIV and 659, Carnap § 43–45.

Against it: Russell *PM* 76, *Einf.* 187, Behmann 29.

Abstraction principle. Russell *Principles* 166
Frege *Grundlagen* 73[D]
Weyl *Handbuch*

Definition of cardinal numbers. Frege *Grundlagen* 789[D], Russell *Principles* 114.

Theory of chains. Begriffsschrift 55

Axiomatic method Weyl 16, further *Einleitung* § 18.

91

Russell, Probleme der Philosophie. Translation by P. Hertz, Erlangen 1926.

p. 60. There are principles a priori that are recognized through intuition (p. 66). Those of logic belong to these, as well as those of induction (empirical induction) and the ethical ones. They are for sure true, without a possibility to prove them from experience. Existential propositions are always empirical.

p. 75 Refutation of the psychological. The law of contradiction is no law of thinking but expresses instead a fact about things (rejection *Kant*).

p. 77 There are things that are neither mental nor corporeal but still exist, relations for example (realm of ideas). He cites Plato explicitly and declares to be in the main in accord with him (p. 78). He finds the rationalists to be right against the empiricists in that there exist universals. (This can be proved because there has to exist, at least, the relation of "similarity.") But they have considered only properties, not relations. Universals are nothing psychological.

p. 86: "Die Welt des Wissens ist ⟦four lines cited⟧

p. 89 A priori knowledge deals exclusively with the relations of universals.

92

p. 90 It must be taken as a fact, discovered by reflecting upon our knowledge, that we have the power of sometimes perceiving such relations between universals, and therefore of sometimes knowing general a priori propositions such as those of arithmetic and logic.[18] (Intuitive knowledge = evident truths = the giving of sense and logical principles (perhaps also ethical principles, later revoked).)

Evidence has grounds but perhaps there exists a highest ground that is the absolute guarantee for truth.

Belief = relation between the one who judges and all of the objects of judgment together (p. 103 ff), for example, (*A* believes that *B* loves *C*). Truth of a belief = existence of the state of affairs that corresponds to it (state of affairs = complex of things in a relation, together with this relation).

Knowledge = firm and true belief that is either given intuitively or that can be inferred in a concrete way from that which is intuitively given. A truth is evident in the highest (absolute) sense if we have acquaintance with the state of affairs to which it corresponds. This can be the case also for universals (p. 118). This evidence is an unconditioned guarantee for truth. However, only the state of affairs possesses this first evidence. The judgment that can be associated to it can possibly reproduce the state of affairs in a false way.

93

Frege, Begriffsschrift 1879

p. 16 Definition of functions:[19] If in an expression a simple or compound sign occurs in one or several places, and if we imagine it to be replaceable at all or also at some of these occurrences, but always with the same one, then we call the part that remains unchanged the function, and that which is replaceable its argument.

Rules of inference:

---

[18] ⟦A literal quote from Russell.⟧

[19] This Fregean definition is improved in "Funktion und Begriff 1891."[D] Difference between the sign for a function and the function.

I $\varphi(x)$, follows $(x)\varphi(x)$

II $A \rightarrow \varphi(x)$, follows $A \rightarrow (x)\varphi(x)$ } p. 21

Axioms:

$a \rightarrow (b \rightarrow a)$

$[c \rightarrow (b \rightarrow a)] \rightarrow [(c \rightarrow b) \rightarrow (c \rightarrow a)]$

$[d \rightarrow (b \rightarrow a)] \rightarrow [b \rightarrow (d \rightarrow a)]$

$(b \rightarrow a) \rightarrow (\sim a \rightarrow \sim b)$

$\sim\sim a \rightarrow a$

$a \rightarrow \sim\sim a$

$c = d \rightarrow [f(a) \rightarrow f(b)]$

$c = c$

$(x)\varphi(x) \rightarrow \varphi(y)$

94

Definition of hereditary p. 58

y follows from *x* in finitely many steps p. 62 and some properties derived

Quarterly Journal 1912 (B 43)[D]

p. 219 *Jourdain*, development of a mathematical theory

(There are sections on Leibniz and Boole in a preceding volume of the *Quart Journ.*, perhaps the latter volume is a continuation)

p. 251 (Nature of the question.) The difficulties that are tied to the concept of a class disappear if one avoids class formation and talks only about concepts and relations, which is possible in the fundamental part of logic.

p. 251 Mentions that Boole's propositional calculus is based on space-time.

p. 288 *Riv. d. Math* $N^{\underline{o}}$ 3 ⟦$N^{\underline{o}}$ 6⟧ 1898, pp. 95–101: Peano emphasizes the difference between $\supset$ and ε (for example, $\sim$ commutes with ε, but not with $\supset$). $\supset$ is transitive but ε not.

H. Grassmann, *Lehrbuch der Arithmetik* 1861, first system that uses the recursive method.

Frege: Die Grundlagen der Arithmetik. Eine logisch-mathematische Untersuchung über den Begriff der Zahl. 1884.[D]

p. 74 ff. Introduction of definition by abstraction

95

Carnap, Logische Aufbau[D]

p. 46 Reducibility (constructibility of an object) to another, each proposition about the first one can be translated to a proposition about the other.

p. 35 Quasi-objects = all except for proper names and propositions = unsaturated expressions

p. 51 two kinds of definition, explicit and for use

p. 57 ff. Extensionality thesis

p. 59 ff. Proof of the extensionality thesis, Fregean distinction between: denotation = what is denoted by a sign, sense = conception or content of thought. Propositional functions that coincide in range have the same denotation but different senses. Apparently non-extensional propositional functions are not propositional functions about their denotation, but about their sense.

(2.) All functions that are considered here are obtained through quantification from pred. Only examples for predicative, with the help of ε and operations of the propositional calculus, but not stated that nothing else exists.

Gödel's references:

Becker, O. (1930) Zur Logik der Modalitäten. *Jahrbuch für Philosophie und phänomenologische Forschung*, vol. 11, pp. 496–548.

Behmann, H. (1922) Beitrage zur Algebra der Logik, insbesondere zum Entscheidungsproblem. *Mathematische Annalen*, vol. 86, pp. 163–229. ⟦13.7.31⟧

Behmann, H. (1927) *Mathematik und Logik.* Mathematisch-physikalische Bibliothek, vol. 71. Leipzig and Berlin 1927, 59 p. ⟦2.3.29⟧

Behmann, H. (1931) Zu den Widersprüchen in der Logik und Mengenlehre. *Jahresbericht der Deutschen Mathematiker-Vereinigung*, vol. 40, pp. 36-48.

Bell. E. (1927) Arithmetic of logic. *Transactions of the American Mathematical Society*, vol. 29, pp. 597–611.

Bergmann, G. (1929) Zur Axiomatik der Elementargeometrie. *Monatshefte für Mathematik und Physik.* vol. 36, pp. 269–284.

Bernays, P. and M. Schönfinkel (1928) Zum Entscheidungsproblem der mathematischen Logik. *Mathematische Annalen*, vol. 99, pp. 342–372. ⟦7.11.30⟧

Bernstein, B. (1916) A set of four independent postulates for Boolean algebras. *Transactions of the American Mathematical Society*, vol. 17, pp. 50–52.

Bernstein, B. (1924a) Complete sets of representations of two-element algebras. *Bulletin of the American Mathematical Society*, vol. 30, pp. 24–30.

Bernstein, B. (1924b) A generalization of the syllogism. *Bulletin of the American Mathematical Society*, vol. 30, pp. 125–127.

Bernstein, B. (1924c) Operations with respect to which the elements of a Boolean algebra form a group. *Transactions of the American Mathematical Society*, vol. 26, pp. 171–175.

Bernstein, B. (1926) On the existence of fields in Boolean algebras. *Transactions of the American Mathematical Society*, vol. 28, pp. 654–657.

Bernstein, B. (1929) Irredundant sets of postulates for the logic of propositions. *Bulletin of the American Mathematical Society*, vol. 35, pp. 545–548.

Bernstein, B. (1931) Whitehead and Russell's theory of deduction as a mathematical science. *Bulletin of the American Mathematical Society*, vol. 37, pp. 480–488.

Bernstein, B. (1932) Note on the condition that a Boolean equation have a unique solution. *American Journal of Mathematics*, vol. 54, pp. 417–418.

Bolzano, B. (1817) *Rein analytischer Beweis des Lehrsatzes, dass zwischen je zwey Werthen, die ein entgegengesetztes Resultat gewähren, wenigstens eine reelle Wurzel der Gleichung liege.* Gottlieb Haase, Prague.

Bolzano, B. (1837) *Wissenschaftslehre*, vols. 1–4.

Bolzano, B. (1920) *Paradoxien des Unendlichen.* Ed. Höfler, with remarks by Hahn.

Borel, E. (1914) *Leçons sur la théorie des fonctions.* 2nd ed.

Boole, G. (1854) *An Investigation of the Laws of Thought.*

Burali-Forti, C. (1897) Questione sui numeri trasfiniti. *Rendiconti di Palermo*, vol. 11, pp. 154–164.

Burali-Forti, C. (1901) Sur les différentes méthodes logiques pour la définition du nombre réel. *Bibliothèque du Congrès International de Philosophie*, vol. 3, pp. 289–307. ⟦3.10.32⟧

Cantor, G. (1872) Über die Ausdehnung eines Satzes aus der Theorie der trigonometrischen Reihen. *Mathematische Annalen*, vol. 5, pp. 123–132.

Cantor, G. (1874) Über eine Eigenschaft des Inbegriffes aller reellen algebraischen Zahlen. *Journal für die reine und angewandte Mathematik*, vol. 77, pp. 258–262.

Cantor, G. (1883) Ueber unendliche, lineare Punktmannigfachkeiten. *Mathematische Annalen*, vol. 21, pp. 546–591.

Carnap, R. (1927) Eigentliche und uneigentliche Begriffe. *Symposion*, vol.1, pp. 355–374. ⟦7.11.28, 11.10.32⟧

Carnap, R. (1928) *Der Logische Aufbau der Welt.* ⟦7.10.32, 21.10.32, 21.11.32, 27.4.33⟧

Carnap, R. (1929) *Abriss der Logistik.*

Church, A. (1925) On irredundant sets of postulates. *Transactions of the American Mathematical Society*, vol. 27, pp. 318–328.

Chwistek, L. (1922) Über die Antinomien der Prinzipien der Mathematik. *Mathematische Zeitschrift*, vol. 14, pp. 236–243.

Chwistek, L. (1922–23) Miara Lebesgue'a. Logiczna analiza i konstrukcya pojkcia miary Lebesgue'a. (Lebesgue measure. Logical analysis and construction of the Lebesgue measure.) *Archiwum Towarzystwa Naukowego we Lwowie, Wydzial III matematyczno-przyrodniczy*, vol. 2, pp. 111–121.

Chwistek, L. (1925) The theory of constructive types (Principles of logic and mathematics). *Annales de la Société Polonaise de Mathématique*, vol. 2, pp. 9–48, and vol. 3, pp. 92–141.

Chwistek, L. (1926) Über die Hypothesen der Mengenlehre. *Mathematische Zeitschrift*, vol. 25, pp. 439–473. ⟦2.4.30, 13.7.31⟧

Chwistek, L. (1929) Neue Grundlagen der Logik und Mathematik. *Mathematische Zeitschrift*, vol. 30, pp. 704–724. ⟦9.3.32⟧

Chwistek, L. (1930) Nouvelles recherches sur les fondements des mathématiques. *Atti del Congresso Internazionale dei Matematici*, vol. 3, pp. 375–380.

Chwistek, L. (1930) Une méthode métamathématique d'analyse. *Comptes-rendus du I Congrès des Mathématiciens des Pays Slaves*, pp. 254–263. ⟦4.4.32⟧

Chwistek, L. (1932) Neue Grundlagen der Logik und Mathematik. Zweite Mitteilung. *Mathematische Zeitschrift*, vol. 34, pp. 527–534.

Clifford, W. (1879) On the types of compound statement involving four classes. *Memoirs of the Literary and Philosophical Society of Manchester*, vol. 6, pp. 81–96.

Couturat, L. (1905) Les définitions en mathématiques. *L'Enseignement mathématique*, vol. 7, pp. 27–40.

Couturat, L. (1906) Pour la logistique (réponse à M. Poincaré). *Revue de métaphysique et de morale*, vol. 14, pp. 208–250. ⟦6.7.32⟧

Couturat, L. (1914) *The Algebra of Logic.*

Curry, H. (1929) An analysis of logical substitution. *American Journal of Mathematics*, vol. 51, pp. 363–384.

Curry, H. (1930) Grundlagen der kombinatorischen Logik. *American Journal of Mathematics*, vol. 52, pp. 509–536 and 789–834.

Curry, H. (1931) The universal quantifier in combinatory logic. *Annals of Mathematics*, vol. 32, pp. 154–180.

Dedekind, R. (1872) *Stetigkeit und irrationale Zahlen.*

Dedekind, R. (1888) *Was sind und sollen die Zahlen?.*

Dines, L. (1914) Complete existential theory of Sheffer's postulates for Boolean algebras. *Bulletin of the American Mathematical Society*, vol. 21, pp. 183–188.

Dixon, A (1906) On "well-ordered" aggregates. *Proceedings of the London Mathematical Society*, vol. 4, pp. 18–20.

Dixon, A. (1907) On a question in the theory of aggregates. *Proceedings of the London Mathematical Society*, vol. 4, pp. 317–319.

Dubislav, W. (1929) Elementarer Nachweis der Widerspruchslosigkeit des Logik-Kalküls. *Journal für die reine und angewandte Mathematik*, vol. 161, pp. 107–112.

Dubislaw, W. (1931) *Die Definition.*

Foster, A. (1932) Formal logic in finite terms. *Annals of Mathematics,* vol. 32, pp. 407–430.

Fraenkel, A. (1922) Der Begriff "definit" und die Unabhängigkeit des Auswahlaxioms. *Sitzungsberichte der Preussischen Akademie der Wissenschaften, Phys.-math. Kl.,* pp. 253–257. ⟦21.10.32⟧

Fraenkel, A. (1925) Untersuchungen über die Grundlagen der Mengenlehre. *Mathematische Zeitschrift,* vol. 22, pp. 250–273.

Fraenkel, A. (1928) *Einleitung in die Mengenlehre.* 2nd ed. ⟦5.4.30⟧

Fraenkel, A. (1930a) Gelöste und ungelöste Probleme im Umkreis des Auswahlprinzips. *Atti del Congresso Internazionale dei Matematici,* vol. 2, pp. 255–259.

Fraenkel, A. (1930) Die heutigen Gegensätze in der Grundlegung der Mathematik. *Erkenntnis,* vol. 1, pp. 286–302.

Frege, G. (1879) *Begriffsschrift, eine nach der arithmetischen nachgebildete Formelsprache des reinen Denkens.*

Frege, G. (1884) *Die Grundlagen der Arithmetik.* ⟦20.11.28, 4.7.32, 21.10.32, 24.11.32 etc to 27.4.33⟧

Frege, G. (1891) *Funktion und Begriff.* Vortrag gehalten in der Sitzung vom 9. Januar 1891 der Jenaischen Gesellschaft für Medicin und Naturwissenschaft. ⟦14.11.32, 28.11.32⟧

Frege, G. (1892) Über Sinn und Bedeutung. *Zeitschrift für Philosophie und philosophische Kritik,* vol. 100, pp. 25–50. ⟦7.10.32⟧

Frege, G. (1892) Über Begriff und Gegenstand. *Vierteljahrsschrift für wissenschaftliche Philosophie,* vol. 16, pp. 192–205. ⟦24.10.32⟧

Frege, G. (1893) *Grundgesetze der Arithmetik,* vol. 1. ⟦11.8.30. 4.7.32⟧

Frege, G. (1903) *Grundgesetze der Arithmetik,* vol. 2. ⟦11.8.30. 14.8.30 Ausleihe, 4.7.32⟧

Frege, G. (1896) Lettera del sig. G. Frege all'Editore. *Rivista di Matematica,* vol. 6 , pp. 53–59.

Frege, G. (1918) Der Gedanke. Eine logische Untersuchung. *Beiträge zur Philosophie der deutschen Idealismus,* vol. 1, pp. 58–77. ⟦7.10.32⟧

Frink, O. (1925) The operations of Boolean algebras. *Annals of Mathematics,* vol. 27, pp. 477–490.

Frink, O. (1928) On the existence of linear algebras in Boolean algebras. *Bulletin of the American Mathematical Society,* vol. 34, pp. 329–333.

Gégalkine, J. (1927) Sur le calcul des propositions dans la logique symbolique. (Russian with French abstract.) *Recueil mathématique de la Société mathématique de Moscou* vol. 34, pp. 9–28.

Gégalkine, J. (1928) L'arithmétisation de la logique symbolique. (Russian with French abstract.) *Recueil mathématique de la Société mathématique de Moscou*, vol. 35, pp. 311–337.

Gégalkine, J. (1929) L'arithmétisation de la logique symbolique. (Russian with French abstract.) *Recueil mathématique de la Société mathématique de Moscou*, vol. 36, pp. 205–338.

Gehman, H. (1926) On irredundant sets of postulates. *Bulletin of the American Mathematical Society*, vol. 32, pp. 159–161.

Gergonne, J. (1816) Essai de dialectique rationelle. *Annales de mathématiques pures et appliquées*, vol. 7, pp. 189–228.

Gerhardt, C. (1890) *Die philosophischen Schriften von Gottfried Wilhelm Leibniz.*

Grassmann, H. (1861) *Lehrbuch der Arithmetik.*

Grassmann, R. (1872) *Die Formenlehre oder Mathematik.*

Grelling, K. and L. Nelson (1908) Bemerkungen zu den Paradoxien von Russell und Burali-Forti. *Abhandlungen der Fries'schen Schule*, vol. 2, pp. 301–334. ⟦19.9.32⟧

Hahn, O. (1909) Zur Axiomatik des logischen Gebietkalkuls. *Archive für systematische Philosophie*, vol. 15, pp. 345–347. ⟦20.7.32⟧

Hahn, O. and O. Neurath (1909) Zum Dualismus in der Logik. *Archive für systematische Philosophie*, vol. 15, pp. 149–162. ⟦20.7.32⟧

Hardy, G. (1929) Mathematical proof. *Mind*, vol. 38, pp. 1–25. ⟦6.7.32⟧

Hartogs, F. Über das Problem der Wohlordnung. *Mathematische Annalen*, vol. 76, pp. 438–443.

Heine, H. (1872) Die Elemente der Funktionenlehre. *Journal für die reine und angewandte Mathematik*, vol. 74, pp. 172–188.

Heiss, R. (1927–8) Der Mechanismus der Paradoxien und das Gesetz der Paradoxienbildung. *Philosophischer Anzeiger*, vol. 2, pp. 403–433. ⟦21.10.32⟧

Hertz, P. (1922) Über Axiomensysteme für beliebige Satzsysteme. *Mathematische Annalen*, vol. 87, pp. 246–269.

Hertz, P. (1923) Über Axiomensysteme für beliebige Satzsysteme. Teil II. *Mathematische Annalen* vol. 89, pp. 76–102.

Hertz, P. (1929) Über Axiomensysteme für beliebige Satzsysteme. *Ma-*

thematische Annalen, vol. 101, pp. 457–514. ⟦9.10.30⟧

Hilbert, D. (1922) Die logischen Grundlagen der Mathematik. *Mathematische Annalen*, vol. 88., pp. 151–166.

Hilbert, D. (1926) Über das Unendliche. *Mathematische Annalen*, vol. 95, pp. 161–190.

Hobson, E. (1907) On the arithmetic continuum. *Proceedings of the London Mathematical Society*, vol. 4, pp. 21–28.

Hölder, O. (1901) Die Axiome der Quantität und die Lehre vom Mass. *Berichte über die Verhandlungen der Königlich Sächsischen Gesellschaft der Wissenschaften zu Leipzig, Mathematisch-Physikaliche Classe*, vol. 53, pp. 1–64.

Hölder, O. (1924) *Die Mathematische Methode.*

Hölder, O. (1926) Der angebliche circulus vitiosus und die sogenannte Grundlagenkrise in der Analysis. *Berichte über die Verhandlungen der Sächsischen Akademie der Wissenschaften zu Leipzig, Mathematisch-physische Klasse*, vol. 78, pp. 243–250.

Hölder, O. (1931) Axiome, empirische Gesetze und mathematische Konstruktionen. *Scientia*, vol. 49, pp. 317–326

Huntington, E. (1902) A complete sets of postulates for the theory of absolute continuous magnitude. *Transactions of the American Mathematical Society*, vol. 3, pp. 264–279.

Huntington, E. (1902) Complete sets of postulates for the theories of positive integral and positive rational numbers. *Transactions of the American Mathematical Society*, vol. 3, pp. 280–284.

Huntington, E. (1904) Sets of independent postulates for the algebra of logic. *Transactions of the American Mathematical Society*, vol. 5, pp. 288–309.

Huntington, E. (1905) *The Continuum as a Type of Order.*

Huntington, E. (1905) A set of postulates for real algebra, comprising postulates for a one-dimensional continuum and for the theory of groups. *Transactions of the American Mathematical Society*, vol. 6, pp. 17–41.

Huntington, E. (1905) A set of postulates for ordinary complex algebra. *Transactions of the American Mathematical Society*, vol. 6, pp. 209–229.

Huntington, E. (1924a) Sets of completely independent postulates for cyclic order. *Proceedings of the National Academy of Sciences*, vol. 10, pp. 74–78.

Huntington, E. (1924b) A new set of postulates for betweenness, with proof of complete independence. *Transactions of the American Mathematical Society*, vol. 23, pp. 276–280.

Huntington, E. and J. Kline (1917) Sets of independent postulates for betweenness. *Transactions of the American Mathematical Society*, vol. 18, pp. 301–325.

Huntington, E. and K. Rosinger (1932) Postulates for separation of point-pairs (reversible order on a closed line). *Proceedings of the American Academy of Arts and Sciences*, vol. 67, pp. 61–145.

Hurwitz, W. (1928) On Bell's arithmetic of Boolean algebra. *Transactions of the American Mathematical Society*, vol. 30, pp. 420–424.

Johnson, W. (1922) *Logic*, vols. I–III. ⟦11.10.32⟧

Jourdain, P. (1912) The development of the theories of mathematical logic and the principles of mathematics. *Quarterly Journal of Mathematics*, vol. 43, pp. 219–314. ⟦23.11.32⟧

Kalmar, L. (1929) Eine Bemerkung zur Entscheidungstheorie. *Acta Scientiarum Mathematicarum Universitatis Szegediensis*, vol. 4, pp. 248–252.

Kaufmann, F. (1931) Bemerkungen zum Grundlagenstreit in Logik und Mathematik. *Erkenntnis*, vol. 2, pp. 262–290.

Kempe, A. (1889–90) On the relation between the logical theory of classes and the geometrical theory of points. *Proceedings of the London Mathematical Society*, vol. 21, pp. 147–182. ⟦11.10.32⟧

Klein, F. (1929) Einige distributive Systeme in Mathematik und Logik. *Jahresbericht der Deutschen Mathematiker-Vereinigung*, vol. 38, pp. 35–40.

Kobrzynski, Z. (1930) Deux types de relations logiques et la méthode de Poretsky. *Annales de la Société Polonaise de Mathematique*, vol. 8, pp. 244–246.

König, J. (1905) Über die Grundlagen der Mengenlehre und das Kontinuumproblem. *Mathematische Annalen* vol. 61, pp. 156–160.

König, J. (1914) *Neue Grundlagen der Logik, Arithmetik und Mengenlehre*. Leipzig.

Korselt, A. (1894) Bemerkungen zur Algebra der Logik. *Mathematische Annalen*, vol. 44, pp. 156–157.

Kossak, E. (1872) *Die Elemente der Arithmetik*.

Kuroda, S. (1931) Zur Algebra der Logik. *Proceedings of the Imperial Academy (of Japan)*, vol. 7, pp. 33–36. ⟦25.11.32⟧

Ladd-Franklin, C. (1883) On the algebra of logic. *Studies in logic by members of the Johns Hopkins University*, Boston, pp. 17–71. ⟦4.7.32⟧

Landau, E. (1930) *Grundlagen der Analysis*.

Langford, C. (1926a) Some theorems on deducibility. *Annals of Mathematics*, vol. 28, pp. 16–40.

Langford, C. (1926b) Analytic completeness of sets of postulates. *Proceedings of the London Mathematical Society*, vol. 25, pp. 115–142.

Langford, C. (1927a) Theorems on deducibility (second paper). *Annals of Mathematics*, vol. 28, pp. 459–471.

Langford, C. (1927b). On inductive relations. *Bulletin of the American Mathematical Society*, vol. 33, pp. 599–607.

Langford, C. (1930) Otherness and dissimilarity. *Mind*, vol. 39, pp. 454–461.

Leśniewski, S. (1931) Grundzüge eines neuen Systems der Grundlagen der Mathematik. *Fundamenta Mathematicae*, vol. 14, pp. 1–81.

Lietzmann, W. (1923) *Trugschlüsse.* Mathematisch-physikalische Bibliothek, vol. 53.

Lindenbaum, A. (1930) Remarques sur une question de la méthode axiomatique. *Fundamenta Mathematicae*, vol. 15, pp. 313–321.

Lindenbaum, A. (1931) Bemerkung zu den vorhergehenden "Bemerkungen . . ." des Herrn J. v. Neumann. *Fundamenta Mathematicae*, vol. 17, pp. 335–336.

Lindenbaum, A. and A. Tarski (1926) Communication sur les recherches de la théorie des ensembles. *Comptes-rendus des séances de la Société des Sciences et des Lettres de Varsovie,Classe III*, vol. 19, pp. 299–330. ⟦25.11.32⟧

Lipschitz, R. (1877) *Lehrbuch der Analysis.*

Löwenheim, L. (1910) Über die Auflösung von Gleichungen im logischen Gebietekalkül. *Mathematische Annalen*, vol. 68, pp. 169–207.

Löwenheim, L. (1913) Über Transformationen im Gebietekalkül. *Mathematische Annalen*, vol. 73, pp. 245–272.

Löwenheim, L. (1915) Über eine Erweiterung des Gebietekalkuls, welche auch die gewöhnliche Algebra umfasst. *Archiv für systematische Philosophie*, vol. 21, pp. 137–148.

Löwenheim, L. (1915) Über Möglichkeiten im Relativkalkül. *Mathematische Annalen*, vol. 76 (1915), pp. 447–470.

Löwenheim, L. (1919) Gebietsdeterminanten. *Mathematische Annalen*, vol. 79, pp. 223–236.

MacColl, H. (1877) The calculus of equivalent statements and integration limits. *Proceedings of the London Mathematical Society*, vol.9, pp. 9–20.

MacColl, H. (1880) Symbolical reasoning. *Mind*, vol. 5, pp. 45–60.

MacColl, H. (1906) *Symbolic logic and its applications.*

McTaggart, J. (1923) Propositions applicable to themselves. *Mind*, vol. 32, pp. 462–464. ⟦6.7.32⟧

Menger, K. (1928) Bemerkungen zu Grundlagenfragen. *Jahresbericht der Deutschen Mathematiker-Vereinigung*, vol. 37, pp. 213–226.

Méray, C. (1872) *Nouveau précis d'Analyse infinitésimale.*

Méray, C. (1894) *Leçons nouvelles sur l'Analyse infinitésimale.*

Miller, J. (1932) Negative Terms in Traditional Logic: Distribution, Immediate Inference and Syllogism. *The Monist*, vol. 42, pp. 96–111.

Mittag-Leffler, G. (1920) Die Zahl: Einleitung zur Theorie der analytischen Funktionen. *Tohoku Mathematical Journal*, vol. 17, pp. 157–209.

Moore, E. (1910) *Introduction to a Form of General Analysis.* ⟦11.7.32⟧

Nagy, A. (1894) Über das Jevons-Clifford'sche Problem. *Monatshefte für Mathematik und Physik*, vol. 5, pp. 331–345.

Nelson, E. (1930) Intensional relations. *Mind*, vol. 39, pp. 440–453. ⟦3.10.32⟧

von Neumann, J. (1928a) Über die Definition durch transfinite Induktion und verwandte Fragen der allgemeinen Mengenlehre. *Mathematische Annalen*, vol. 99, pp. 373–391. ⟦7.11.30⟧

von Neumann, J. (1928b) Die Axiomatisierung der Mengenlehre. *Mathematische Zeitschrift*, vol. 27 , pp. 669–752. ⟦26.10.30⟧

von Neumann, J. (1931) Bemerkungen zu den Ausführungen von Herrn St. Lesniewski über meine Arbeit"Zur Hilbertschen Beweistheorie." *Fundamenta Mathematicae*, vol. 17, pp. 331–334.

Nicod, J. (1916) A reduction in the number of primitive propositions of logic. *Proceedings of the Cambridge Philosophical Society*, vol. 19, pp. 32–41. ⟦9.7.31, 23.10.31, 14.11.31⟧

Orloff, I. (1928) Sur la théorie de la compatabilité des propositions. Russian with French abstract. *Recueil mathématique de la Société mathématique de Moscou*, vol. 35, pp. 263–286.

Padoa, A. (1901) Essai d'une théorie algébrique des nombres entiers, précédé d'une introduction logique a une théorie déductive quelconque. *Bibliothèque du Congrès International de Philosophie*, vol. 3, pp. 309–365. ⟦3.10.32⟧

Pasch, M. (1909) *Grundlagen der Analysis.*

Pasch, M. (1914) *Veränderliche und Funktion.*

Pasch, M. (1919) *Mathematik und Logik.* ⟦6.7.32⟧

Pasch, M. (1927) *Mathematik am Ursprung: Gesammelte Abhandlungen.* ⟦6.7.32⟧

Peano, G. (1889) *Arithmetices principia, nova methodo exposita.* ⟦3.11.32⟧

Peano, G. (1895) *Formulaire de mathématiques,* vol. 1. ⟦4.7.32, 9.9.32, 20.10.32⟧

Peano, G. (1897) *Formulaire de mathématiques,* vol. 2 § 1, *Logique mathématique.*

Peano, G. (1898) *Formulaire de mathématiques,* vol. 2 § 2, *Arithmétique.*

Peano, G. (1899) *Formulaire de mathématiques,* vol. 2 § 3, *Logique mathématique–Arithmétique–Limites–Nombres complexes–Vecterus–Dérivéees–Integrales.*

Peano, G. (1901) *Formulaire de mathématiques,* vol. 3.

Peano, G. (1901) Les définitions mathématiques. *Bibliothèque du Congrès International de Philosophie,* vol. 3, pp. 279–288. ⟦3.10.32⟧

Peano, G. (1902–3) *Formulaire mathématique,* vol. 4.

Peano, G. (1905–8) *Formulario mathematico,* vol. 5.

Peano, G. (1906) Addizione. *Revista de Mathematica,* vol 8, pp. 143–157.

Peirce, C. (1867) On an improvement in Boole's calculus of logic. *Proceedings of the American Academy of Arts and Sciences,* vol. 7, pp. 250–261.

Peirce, C. (1867) Upon the logic of mathematics. *Proceedings of the American Academy of Arts and Sciences,* vol. 7, pp. 402–412.

Peirce, C. (1870) Description of a notation for the logic of relatives, resulting from an amplification of the conceptions of Boole's calculus of logic. *Memoirs of the American Academy of Arts and Sciences,* vol. 9, pp. 317–378. ⟦4.7.32⟧

Poincaré, H. (1904) *Wissenschaft und Hypothese.* ⟦11.7.32⟧

Poincaré, H. (1905) Les mathématiques et la logique. *Revue de métaphysique et de morale,* vol. 13, pp. 815–835. ⟦6.7.32⟧

Poincaré, H. (1906a) Les mathématiques et la logique. *Revue de métaphysique et de morale,* vol. 14, pp. 17–34. ⟦18.12.29, 6.7.32⟧

Poincaré, H. (1906b) Les mathématiques et la logique (suite et fin). *Revue de métaphysique et de morale,* vol. 14, pp. 294–317. ⟦18.12.29, 6.7.32⟧

Poincaré, H. (1909a) La logique de l'infini. *Revue de métaphysique et de morale,* vol. 17, 461–482.

Poincaré, H. (1909b) Réflexions sur les deux notes précédentes. *Acta mathematica,* vol. 32, pp. 195–200.

Poincaré, H. (1912) La logique de l'infini. *Scientia,* vol. 12, pp. 1–11.

⟦21.12.28⟧

Poincaré, H. (1913) *Letzte Gedanken.*

Poretzky, P. (1898–9) Sept lois fondamentales de la théorie des égalités logiques. *Bulletin de la Société Physico-Mathématique de Kasan*, pp. 33–103, 129–181, 183–216.

Post, E. (1921a) Introduction to a general theory of elementary propositions. *American Journal of Mathematics*, vol. 43, pp. 163–185. ⟦8.7.31⟧

Post, E. (1921b) On a simple class of deductive systems. *Bulletin of the American Mathematical Society*, vol. 27, pp. 396–397.

Ramsey, F. (1926) The foundations of mathematics. *Proceedings of the London Mathematical Society*, vol. 25, pp. 338–384.

Ramsey, F. (1926–7) Mathematical logic. *The Mathematical Gazette*, vol. 13, pp. 185–194.

Ramsey, F. (1929) On a problem of formal logic. *Proceedings of the London Mathematical Society*, vol. 30, pp. 264–286.

Richard, J. (1903) *Sur la philosophie des mathématiques.*

Richard, J. (1906) Lettre à Monsieur le rédacteur de la Revue Générale des Sciences. *Acta Mathematica*, vol. 30, pp. 295–296. ⟦6.7.32⟧

Richard, J. (1907) Sur un paradoxe de la théorie des ensembles et sur l'axiome Zermelo. *L'Enseignement mathématique*, vol. 9, pp. 94–98.

Richard, J. (1920) Considérations sur la logique et les ensembles. *Revue de métaphysique et de morale*, vol. 27, pp. 355–369.

Royce, J. (1905) The relations of the principles of logic to the foundations of geometry. *Transactions of the American Mathematical Society*, vol. 6, pp. 535–415.

Royce, J. (1912) Die Prinzipien der Logik. *Encyclopädie der philosophischen Wissenschaften*, vol. 1, pp. 61–135. ⟦24.10.32⟧

Russell, B. (1903) *The Principles of Mathematics.*

Russell, B. (1905) On denoting. *Mind*, vol. 14, pp. 479–493.

Russell, B. (1906a) The theory of implication. *American Journal of Mathematics*, vol. 28, pp. 159–202.

Russell, B. (1906b) Les paradoxes de la logique. *Revue de métaphysique et de morale*, vol. 14, pp. 627–650. ⟦18.12.29, 6.7.32⟧

Russell, B. (1906c) On some difficulties in the theory of transfinite numbers and order types. *Proceedings of the London Mathematical Society*, vol. 4, pp. 29–53.

Russell, B. (1908) Mathematical logic as based on the theory of types. *American Journal of Mathematics*, vol. 18, pp. 222–262.

Russell, B. (1910) La théorie des types logiques. *Revue de Métaphysique et de Morale*, vol. 18, pp. 263–301.

Russell, B. (1923) *Einführung in die mathematische Philosophie.*

Russell, B. (1926) *Unser Wissen von der Aussenwelt.*

Russell, B. (1929) *Die Philosophie der Materie.* Übersetzt von Kurt Grelling.

Schönfinkel, M. (1924) Über die Bausteine der mathematischen Logik. *Mathematische Annalen*, vol. 92, pp. 305–316.

Schröder, E. (1890–95) *Vorlesungen über die Algebra der Logik.* Vols. 1–3. ⟦3.11.28, 24.10.30⟧

Sheffer, H. (1913) A set of five independent postulates for Boolean algebras, with application to logical constants. *Transactions of the American Mathematical Society*, vol. 14, pp. 481–488. ⟦23.10.31⟧

Sheffer, H. (1916) Mutually prime postulates. *Bulletin of the American Mathematical Society*, vol. 1, p. 287.

Sheffer, H. (1927) Notational relativity. *Proceedings of the Sixth International Congress of Philosophy, Harvard University*, pp. 348–351. ⟦3.10.32⟧

Skolem, T. (1920) Logisch-kombinatorische Untersuchungen über die Erfüllbarkeit oder Beweisbarkeit mathematischer Sätze nebst einem Theoreme über dichte Mengen. *Skrifter utgit av Videnskapsselskapet i Kristiania, I. Matematisk-naturvidenskabelig klasse 1920*, no. 4.

Skolem, T. (1923) Bemerkungen zur axiomatischen Begründung der Mengenlehre. *Wissenschaftliche Vorträge gehalten auf dem Fünften Kongress der Skandinavischen Mathematiker in Helsingfors*, pp. 217–232. ⟦7.11.28, 12.9.30⟧

Skolem, T. (1929) Über einige Grundlagenfragen der Mathematik. *Skrifter utgitt av Det Norske Videnskaps-Akademi i Oslo, I. Matematisk-naturvidenskapelig klasse*, 1929, no. 4, 49 p. ⟦11.5.30, 12.9.30⟧

Stäkel, P. (1907) Zu H. Webers Elementarer Mengenlehre. *Jahresbericht der Deutschen Mathematiker-Vereinigung*, vol. 16, pp. 425–428.

Stamm, E. (1911) Beitrag zur Algebra der Logik. *Monatshefte für Mathematik und Physik*, vol. 22, pp. 137–149.

Stamm, E. (1922) Connexione inter operationes arithmetica et logica. *Annales de la Société Polonaise de Mathematique*, vol. 1, pp. 58–69.

Stolz, O. (1885) *Vorlesungen über allgemeine Arithmetik.*

Study, E. (1929) Die angeblichen Antinomien der Mengenlehre. *Sitzungs-*

berichte der preussischen Akademie der Wissenschaften, physikalisch-mathematische Klasse. vol. 19, pp. 255–267.

Tagaki, T. (1931) Zur Theorie der natürlichen Zahlen. *Proceedings of the imperial Academy of Japan*, vol. 7, pp. 29–30. ⟦17.8.32⟧

Tajtelbaum, A. (1923) Sur le terme primitif de la logistique. *Fundamenta Mathematicae*, vol. 4, pp. 196–200.

Tajtelbaum-Tarski, A. (1924) Sur les truth-functions au sens de MM. Russell et Whitehead. *Fundamenta Mathematicae*, vol. 5, pp. 59–74.

Tannery, J. (1894) *Leçons d'arithmétique*.

Tarski, A. (1924) Sur les ensembles finis. *Fundamenta Mathematicae*, vol. 6, pp. 45–95.

Tarski, A. (1930) Fundamentale Begriffe der Methodologie der deduktiven Wissenschaften I. *Monatshefte für Mathematik und Physik* , vol. 37, pp. 361–404. ⟦9.3.32⟧

Taylor., J. (1917) Complete existential theory of Bernstein's set of four postulates for Boolean algebras. *Annals of Mathematics*, vol. 19, pp. 64–69.

Veblen, O. (1904) A system of axioms for geometry. *Transactions of the American Mathematical Society*, vol. 5, pp. 343–384.

Veblen, O. (1906) Review of Huntington: *The Continuum as a Type of Order*, (1905). *Bulletin of the American Mathematical Society*, vol. 12, pp. 302–305.

Venn, J. (1881) *Symbolic Logic*. Second ed. 1894. ⟦14.7.32⟧

Weiss, P. (1932) The metaphysics and logic of classes. *The Monist*, vol. 42, pp. 112–154.

Wernick, G. (1929) Die Unabhängigkeit des zweiten distributiven Gesetzes von den übrigen Axiomen der Logistik. *Journal für die reine und angewandte Mathematik*, vol 161, pp. 123–134.

Weyl, H. (1918) *Das Kontinuum: Kritische Untersuchungen über die Grundlagen der Analysis*.

Weyl, H. (1919) Der circulus vitiosus in der heutigen Begründung der Analysis. *Jahresbericht der Deutschen Mathematiker-Vereinigung*, vol. 28, pp. 85–92.

Weyl, H. (1925) Die heutige Erkenntnislage in der Mathematik. *Symposion*, vol. 1, pp. 1–32. ⟦7.11.28, 11.10.32⟧

Weyl, H. (1926) *Philosophie der Mathematik und Naturwissenschaft*.

Whitehead, A. (1898) *A Treatise on Universal Algebra with Applications*.

Whitehead, A. (1901) Memoir on the algebra of symbolic logic. *American*

Journal of Mathematics, vol. 23, pp. 139–165, 297–316.

Whitehead, A. (1902) On cardinal numbers. *American Journal of Mathematics*, vol. 24, pp. 367–394.

Whitehead, A. (1903) The logic of relations, logical substitution groups, and cardinal numbers. *American Journal of Mathematics*, vol. 25, pp. 157–178.

Whitehead, A. and B. Russell (1925–27) *Principia Mathematica*. Vols. 1–3, 2nd ed.

Wiener, N. (1917) Certain formal invariances in Boolean algebras. *Transactions of the American Mathematical Society*, vol. 18, pp. 65–72.

Wiener, N. (1919–20) A new theory of measurement: A study in the logic of mathematics. *Proceedings of the London Mathematical Society*, vol. 19, pp. 181–205.

Wilson, E. (1908) Logic and the continuum. *Bulletin of the American Mathematical Society*, vol. 14, pp. 432–443.

Wittgenstein, L. (1921) Logisch-philosophische Abhandlung. With an introduction (in German) by Russell. *Annalen der Naturphilosophie*, vol. 14, pp. 185–262.

Yule, D. (1926) Zur Grundlegung des Klassenkalküls. *Mathematische Annalen*, vol. 95, pp. 446–452.

Zaremba, S. (1916) Essai sur la théorie de la démonstration dans les sciences mathématiques. *L'Enseignement Mathématique*, vol. 18, pp. 5–44.

Zermelo, E. (1908) Neuer Beweis für die Möglichkeit einer Wohlordnung. *Mathematische Annalen*, vol. 65, pp. 107–128.

Zylinski, E. (1925) Some remarks concerning the theory of deduction. *Fundamenta Mathematicae*, vol. 7, pp. 203–209. ⟦11.7.32⟧

Index of names for Gödel's chapters:[20]

[20] ⟦Numbers for frames from reel 24, in the order in which they occur in the main text.⟧

Index of names for Gödel's notebooks of excerpts:[21]

[21]⟦Numbers with an asterisk refer to the untitled notebook and come first in order.⟧

MIX
Papier aus verantwortungsvollen Quellen
Paper from responsible sources
FSC® C105338

If you have any concerns about our products,
you can contact us on
ProductSafety@springernature.com

In case Publisher is established outside the EU,
the EU authorized representative is:
Springer Nature Customer Service Center GmbH
Europaplatz 3, 69115 Heidelberg, Germany

Printed by Libri Plureos GmbH
in Hamburg, Germany